Outside the Anthropological Machine

In the midst of the climate crisis and the threat of the sixth extinction, we can no longer claim to be the masters of nature. Rather, we need to unlearn our species' arrogance for the sake of all animals, human and non-human. Rethinking our being-in-the-world as *Homo sapiens*, this monograph argues, starts precisely from the way we relate to our closer companion species. The authors gathered here endeavour to find multiple exit strategies from the anthropocentric paradigms that have bound the human and social sciences. Part I investigates the unexplored margins of human history by re-reading historical events, literary texts, and scientific findings from an animal's perspective, rather than a human's. Part II explores different forms of human-animal relationships, putting the emphasis on the institutions, spaces, and discourses that frame our interactions with animals. Part III engages with processes of "translation" that aim to render animals' experience and perception into human words and visual language.

Chiara Mengozzi is a researcher at Charles University in Prague (Faculty of Arts), where she teaches literary theory, as well as Italian and French contemporary literatures. She is also Associated Researcher at CEFRES in Prague (French Research Center in Social Sciences). Her research focuses on animal studies, ecocriticism, climate fiction, postcolonial and migrant literatures, world literature, and translation and reception theory. She is the author of an Italian monograph entitled *Narrazioni contese. Vent'anni di scritture italiane della migrazione* (2013). She is currently investigating the relation between science and literature in the context of climate change.

Perspectives on the Non-Human in Literature and Culture
Series Editor: Karen Raber
University of Mississippi, USA

Literary and cultural criticism has ventured into a brave new world in recent decades: posthumanism, ecocriticism, critical animal studies, the new materialisms, the new vitalism, and other related approaches have transformed the critical environment, reinvigorating our encounters with familiar texts, and inviting us to take note of new or neglected ones. A vast array of non-human creatures, things, and forces are now emerging as important agents in their own right. Inspired by human concern for an ailing planet, ecocriticism has grappled with the question of how important works of art can be to the preservation of something we have traditionally called "nature." Yet literature's capacity to take us on unexpected journeys through the networks of affiliation and affinity we share with the earth on which we dwell—and without which we die—and to confront us with the drama of our common struggle to survive and thrive has not diminished in the face of what Lyn White Jr. called "our ecological crisis." From animals to androids, non-human creatures and objects populate critical analyses in increasingly complex ways, complicating our conception of the cosmos by dethroning the individual subject and dismantling the comfortable categories through which we have interpreted our existence. Until now, however, the elements that compose this wave of scholarship on non-human entities have had limited places to gather to be nurtured as a collective project. "Perspectives on the Non-Human in Literature and Culture" provides that local habitation. In this series, readers will find creatures of all descriptions, as well as every other form of biological life; they will also meet the non-biological, the microscopic, the ethereal, the intangible. It is our goal for the series to provide an encounter zone where all forms of human engagement with the non-human in all periods and national literatures can be explored, and where the discoveries that result can speak to one another, as well as to scholars and students.

Writing the Land, Writing Humanity
The Maya Literary Renaissance
Charles Pigott

Outside the Anthropological Machine
Crossing the Human-Animal Divide and Other Exit Strategies
Chiara Mengozzi

For more information about this series, please visit: www.routledge.com/Perspectives-on-the-Non-Human-in-Literature-and-Culture/book-series/PNHLC

Outside the Anthropological Machine

Crossing the Human-Animal Divide and Other Exit Strategies

Edited by
Chiara Mengozzi

Routledge
Taylor & Francis Group
NEW YORK AND LONDON

First published 2021
by Routledge
52 Vanderbilt Avenue, New York, NY 10017

and by Routledge
2 Park Square, Milton Park, Abingdon, Oxon, OX14 4RN

Routledge is an imprint of the Taylor & Francis Group, an informa business

Library of Congress Cataloging-in-Publication Data
A catalog record for this book has been requested

ISBN: 9780367504441 (hbk)
ISBN: 9781003049883 (ebk)

Typeset in Sabon
by Apex CoVantage, LLC

This work was supported by the European Regional Development Fund project "Creativity and Adaptability as Conditions of the Success of Europe in an Interrelated World" (reg. no.: CZ.02.1.01/0.0/0.0/16_019/0000734).

EUROPEAN UNION
European Structural and Investment Funds
Operational Programme Research,
Development and Education

MINISTRY OF EDUCATION,
YOUTH AND SPORTS

Contents

Introduction

Ways Out of the Anthropological Machine, or How and Why to Defamiliarize Ourselves

Chiara Mengozzi

In his seminal book entitled *The Open: Man and Animal*, Giorgio Agamben qualifies the anthropological machine as "the device for producing the recognition of the human" (2004, 26) through the articulation of a fundamental dichotomy that separates man and animal, the latter being defined in a privative way (as lacking rationality, language, and self-consciousness, unable to laugh, poor-in-world, etc.). This arbitrary division—as Agamben warns—is likely to produce dangerous zones of indeterminacy, historically occupied by certain ethnic or social groups, who have filled the so-called missing link between the "true" human being and the abjected animal. In other words, the separation of man and animal is in itself a division within the category of man: in ancient times, the machine would function through the humanization of the animal (the slave, the barbarian, the *enfant sauvage*, and the foreigner were all animal figures in human form), while its modern version has worked through the exclusion of an inside, that is, by animalizing the human (such as the Jews during the Holocaust). Two points, however, remain unexplored in Agamben's inquiry. Firstly, Agamben does not describe both sides of the anthropological machine, neglecting what consequences it implies for the life of animals:[1] after all, the animalization of racially marked human groups and the dehumanization of the enemy prove to be an effective instrument of oppression, subjugation, and extermination precisely because the dominion over animals and the violence against them are left unquestioned by the humanist schema of thought and do not constitute an ethical issue. Secondly, Agamben repeatedly asserts in his book that the machine must be stopped, for ethical and political reasons, but he does not indicate what strategies one needs to adopt in order to render the machine inoperative, even though, admittedly, understanding how it works is the first step towards being able to stop it.

This collective monograph follows the post-anthropocentric path opened by Agamben by addressing the lacunae previously mentioned. On the one hand, the authors examine how the anthropological machine affects the lives of animals, while preventing us from seeing and acknowledging to what extent our history, science, language, and everyday practices are embedded in and influenced by inter-species relations. On the other hand, they search for various exit strategies from the anthropocentric paradigm to include animals' material existences, inner lives, and points of view within the orbit of human thought and imagination, however difficult, partial, and counter-intuitive this attempt may seem.

Although Agamben does not help us much when it comes to providing a way out of this "performative apparatus" (Calarco 2015, 54), he does give some hints when claiming that

> to render inoperative the machine that governs our conception of man will therefore mean no longer to seek new—more effective or more authentic—articulations, but rather to show the central emptiness, the hiatus that—within man—separates man and animal, and to risk ourselves in this emptiness: the suspension of the suspension, Shabbat of both animal and man.
>
> (Agamben 2004, 92)

One might object that the reference to the Shabbat of the Jewish tradition (etymologically "rest", "ceasing from work") is a vague suggestion, and ultimately quite suspect, given that the three major monotheistic religions, far from questioning the anthropological divide, played a significant role in establishing the man-animal binary that allows the machine to function. It would be unfair, however, to dismiss Agamben's reference to religion so quickly. By bringing into play a third term (i.e. the "divine"), he is implicitly pointing out that the dyad human-animal might be insufficient to understand (and modify) human-animal relations and, therefore, that it should be abandoned, or at least complicated. However fruitful it may be to follow Agamben on this path, this monograph rather engages in a more secular approach, leaving aside the sphere of the religious. Instead, the inspiration for the present monograph is his proposal to risk ourselves in the emptiness, the "hiatus . . . that separates man and animal." Since Agamben is mostly concerned with the anthropological machine's dangerous effects on the human side, he promptly adds in a parenthetical statement that the hiatus he is referring to is first and foremost the one "within man," but his exhortation can also act as a spur towards new ways of thinking about both human and animal life beyond the dichotomic opposition. Drawing from Walter Benjamin's conception of the relationship between humans and nature, Agamben states that the two terms should not enter into a relation of mastery (man over nature or, vice-versa, nature over man) nor be surpassed through their dialectical synthesis, but rather they should be reciprocally suspended in order to unlearn these traditional ways of knowing and thinking and to risk ourselves in the interval between man and nature, between human and animal life: "What is decisive here—he continues—is only the 'between', the interval or, we might say, the play between the two terms, their immediate constellation in a non-coincidence" (Agamben 2004, 83). But what is the nature of this "between," and how is it to be understood? In other words, to put out of action the "othering machine"—whereby the human is re-produced against an allegedly animal nature which would embody man's inverted mirror—should we insist on "the indistinction of human and animal life prior to their separation" (Calarco 2015, 54), as suggested by Agamben at the end of the book, or should we rather turn towards different and more complex ways to articulate the separation, as proposed by Derrida? These two approaches are often (and understandably) seen as diametrically opposed, if not incompatible, but it would probably be more correct to think of them in terms of necessary co-implication. If we start our reasoning by stressing that there is no "single, linear, indivisible, oppositional limit" that would "allow us to cut once and for all where we would in general like to cut" (Derrida 1995, 285), it is certainly to acknowledge the plentiful variety, richness, and singularity of both animal and human worlds but also to put both humans and non-humans in a broader web of relations beyond the inherited dichotomies (nature/culture, subject/object, mind/body). Here, we recognize that there exist multiple zones of contact, overlap, and indeterminacy, since what we thought of as the proper of man (rationality, agency, language, etc.) is or might be shared by numerous animal species. Vice-versa, if we shift the focus onto our shared embodiment and vulnerability, the "secret bond that ties man to life" (Agamben 2004, 83) and what Jean-Christophe Bailly has called "le fonds commun d'expérience," it is not to erase or disregard the "irreducible difference that animals may represent for us" (Weil 2012, 20) but to stress that each species reaches individuation through co-evolution and interactions, to acknowledge that "what distinguished men from animals was born of their relationship with them" (Berger 1992, 9), and, ultimately, to allow "new groupings and differences to emerge" (Calarco 2015, 56). In other words, these two approaches, notwithstanding their antithetical starting points, end up involving, or at least partially contemplating, the other perspective as an indispensable counterpart.

We can then return to the issue regarding the proper way to understand the interval between human and non-human life and answer as follows: just as the hyphen is simultaneously a rift and an enjambment (i.e. it blends and divides the two terms all at once) in hyphenated words, to think the interstitial space between human and non-human life probably means to fully embrace the aporia of (in)differentiation, to keep it alive, so to speak, without seeking to resolve the impasse by means of dialectical thinking or through the unilateral choice of one of the two "forking paths." But the problem is that it is not viable, after all, to remain (and write) in the condition of a permanent or protracted aporia. That is why the authors of the monographs—starting from the common assumptions that animals are simultaneously like us and different from us (but each of them in their particular way) and that their life is neither inferior nor superior than humans' in terms of intensity, plenitude, or intrinsic value—will alternatively insist more on one of the two perspectives, keeping always in the background the other, as a sort of reminder, a necessary supplement.

What I would like to emphasize, however, is rather another point. Whether we start by risking ourselves in the open chain of "*différance*" or, conversely, by positing the "indistinction" as prior and more fundamental, what is at stake in these two ways for undoing the traditional boundary between human and animal is nothing less than the opportunity to embark in a difficult but rewarding venture of self-displacement and defamiliarization—this latter term translating the concept of "*ostranenie*," coined by Viktor Shklovsky in a deservedly famous essay called "Art as Device." Against the automatism of perception and the familiar understanding that covers rather than discloses things, we need to make them "unfamiliar" and to undertake the imaginary experience of looking at them as for the first time, with new eyes, from another point of view—it is no coincidence that Shklovsky had chosen the example of Tolstoy's short story about the horse Kholstomer to illustrate how literature can reactivate our perception. Shklovsky, indeed, aims to point out the cognitive value of literature and its formal devices, whereby objects and everyday practices can be revitalized. However, while it is true that literature does have the special power to recover "the sensation of life [. . .] to make one feel things [. . .] to make the stone stony" (2017, 80)—as Shklovsky puts it—one should refrain from attributing the effect of defamiliarization to literature in its entirety, if only because it frequently works the other way around, bringing closer to us people, experiences, worlds that seem radically different from our own. Moreover, it must be admitted that literature is not the only discursive resource we have at our disposal to unsettle and therefore to intensify, deepen, and widen our knowing and perception processes. What I am trying to suggest here is that, in order to step outside the anthropological machine, a twofold and joint path must be undertaken. Not only do we need to shed a new light on animals, "learning to learn" from all the living beings with which we share the planet, but we are also urged to undermine our taken-for-granted identities and to unlearn our alleged exceptionalism and privileges. This is a task to which all human and social sciences can (and should) contribute, each of them by its own means and in dialogue with other disciplinary fields, such as ethology, neurosciences, evolutionary biology, and so on. The aim of this monograph is thus to push us to see and to apprehend the familiar in an unfamiliar way and vice-versa, exploring sundry strategies which allow us to radically reconsider ourselves and our being-in-the-world, because only by assuming that "the human [might] no longer [be] the center or the chief point of reference" (Calarco 2015, 56) can we hope to let animals' lives and worlds emerge from the background to which they have been relegated, especially in hyper-industrialized societies.

These strategies revolve around three major challenges that must be tackled once we have decided that the anthropological machine must be stopped for the sake of both human and non-human animals. First, what is at issue is the effort to place us at the margins of the (hi)story, while bringing other co-protagonists into the foreground; the second step consists in shifting the focus onto the human-animal relation (be it concrete or metaphorical) and what allows, mediates, or prevents a genuine encounter; and third, it is about seeking a path for us to come out of ourselves in order to venture into the territory of other animals, that is, into their perceptual and cognitive worlds.

Before giving insight into the content of the three parts of the volume, or detailing how the authors engage in different strategies of (de)familiarization, the question remains: why is it worth partaking in these demanding post- and anti-anthropocentric exercises? The reasons are manifold, but let me mention at least three of them.

In his influential essay "Why Look at Animals?," John Berger, while speaking about the strong bond that used to tie humans and animals in ancient times, recalls that

> such an unspeaking companionship was felt to be so equal that often one finds the conviction that it was man who lacked the capacity to speak with animals—hence the stories and legends of exceptional beings like Orpheus who could talk with animals in their own language.
>
> (1992, 6)

What, in Berger's formulation, is a sort of Eden-like condition, which seems irrecoverably lost in modern times, becomes in Jean-Christophe Bailly's *Le Parti pris des animaux* the principal duty and responsibility of modern thinkers and artists: "animals, almost by definition, do not speak. . . . Besides, it is not about giving them a voice by coming closer to them . . . it is about going one step ahead of their silence and identifying what is said there" (Bailly 2013, transl. mine).[2] In other words, if we assume that animals are not poor- but rich-in-world, that their silence is not the sign of deprivation but littered with traces, postures, movements, perceptions, cries, squeaks, affections, and abilities, and if we acknowledge that they might see, feel, and touch something that eludes us, then we must also admit that confining ourselves to the human side of things—without even trying to translate into human language those non-human lives and worlds—means precluding the opportunity to escape "from our spiritual narrowness" (Bailly 2013, transl. mine). As Kari Weil poignantly wrote,

> we may know animals in ways they cannot—we may know their breeds, their colour, their weight, their names, their "histories"—but

> they may also know us in ways that we cannot know because they know the world and us by other means.
>
> (Weil 2012, 11)

The purpose then is not to replace one point of view with another but to enrich the apprehension of our common world, as all the authors gathered herein aim to demonstrate. Indeed, they resonate with the common aim of opening to (and through) animals our ways of thinking and knowing, not because "it would be amusing, but because at the end of each of these pathways, our vision of the landscape [and—we may add—of history, every day practises, literature, and the whole planet] emerges enlarged, enriched, emancipated" (Bailly 2013, transl. mine).[3]

Besides that, there is another reason for displacing the human subject from our studies and attempting forays into the multiplicity and singularity of animal worlds and points of view. These endeavours have the potential to renew the so-called human and social sciences, to stimulate self-reflexivity in our disciplines, and to encourage cross-disciplinary approaches. Yet pursuing a non-anthropocentric line of inquiry should not simply add a new object of study, until recently overlooked or considered as auxiliary, to the list of academic topics, but should rather lead to a search for conceptual and methodological frameworks that allow this cognitive shift. For this reason, all the chapters of the volume engage, to a greater or lesser extent, with the limits of our disciplinary protocols, methods, and assumptions. Indeed, it is neither appropriate nor possible to include animals in historical research without challenging the traditional distinction between natural and human history and deeply reappraising the very interpretation of historical sources, nor to describe our everyday practices in the light of the inter-species bonds without broadening the notion of subject, individual, and agency, along with exploring the frontiers and potentialities of verbal and non-verbal language when striving to imagine and represent what it is like to be a non-human animal. Therefore, making our disciplines less anthropocentric implies several journeys back and forth along three paths. It means pushing our disciplines out of their anthropocentric frame, exploring how and to what extent they can deepen our understanding of non-human life, even taking the opposite direction to inquire what animals "have to say" about our disciplines (i.e. how the inclusion of animals allows the reshaping of our concepts), while simultaneously taking into account other fields (such as natural sciences, neurosciences, ethology, mesology, etc.), to find what, outside the human sciences, can shake up our current habits and assumptions.

The final reason to displace the human subject is that, in the midst of the climate crisis, "we can no longer enjoy a carefree anthropocentrism," as Laura Pugno, a contemporary Italian writer, has put it.[4] Pursuing an anti- and post-anthropocentric line of research, then, is not simply an act of generosity towards other living beings but has become a matter of survival for our own species.

> The climate crisis is about waking up to the rude shock of the planet's otherness. The planet, to speak with Spivak again, "is in the species of alterity, belonging to another system." And "yet," as she puts it, "we inhabit it." If there is to be a comprehensive politics of climate change, it has to begin from this perspective. The realization that humans—all humans, rich or poor—come late in the planet's life and dwell more in the position of passing guests than possessive hosts.
> (Chakrabarty 2014, 23)

In other words, climate change—as the most conspicuous effect of the epoch known as Anthropocene, together with what is called by biologists "the sixth mass extinction"—opens a new twofold awareness. On the one hand, it blatantly demonstrates that what we used to call "nature" does not represent an inert background that can be transformed at will without consequences and urges us to acknowledge the uncanny alterity of non-human agents with which we need to share and constitute a common world. As is worth recalling, not only is this world composed mostly of "not-us" (animals, plants, bacteria, microbes, not to mention the inanimate world) but also the frontier between us and "not-us" is becoming more and more porous (see Morton 2018). Thus, limiting ourselves to the human side of things means precluding the possibility to understand what is happening on our planet and giving up the opportunity to change the course of the events. On the other hand, the current environmental crisis makes us realize that an anthropocentric approach and conduct, both in our research and in our lives, are untenable and ultimately noxious first and foremost for human beings. It is worth asking, in fact, whether a world without (or with much less) wild animals would still be a human world. Certainly, it would be a world haunted by a deep melancholia for this irreparable loss for which we are responsible, but it would also be a world made far less liveable for human beings, considering how certain species (bees, to give just one telling example) are vitally important to the health of the planet and its ecological balances. In an interesting article entitled "Beyond Anthropocentrism in Historical Studies," Ewa Domanska rightly asked herself what kind of humanities we need today: "I might answer that we need the kind of knowledge, cognition, and human science that have survival value and might help in the protection and continuation of the species" (2011, 120). Of course, it is not a question of resolving massive problems such as the environmental crisis and the ruinous collapse of biodiversity through the humanities, but it would be equally simplistic to think that it can be done without the contribution of the humanities at large, if nothing because, before acting responsibly (and being persuaded to do so), a radical rethinking of our being-in-the-world as *homo sapiens* is necessary, and this begins from the ways we look at our closer companion species. Paradoxically, it is precisely from the sense of the impending catastrophe that this new vision and awareness arise, such as when something breaks or a mechanism gets stuck and the object, suddenly removed from the automatism of its use, not only becomes visible to us again but also manifests itself in a new light. Yet how can we fashion this sense of the imminent (or rather "ongoing") disaster into a key that allows us to displace the paradigms that have upset planetary balances, and with which we used to interpret our being-in-the-world? Responding to catastrophic events with other catastrophic or alarmist narratives is neither the only nor always the best way to take on the challenges posed by the current ecological crisis. That is why the chapters of this monograph prefer to question the anthropological machine, to understand its functioning and effects on the lives of animals, in order to stimulate a different understanding of our *Umwelt*, understood as a space of coexistence, made of encounters and symbioses (or vice-versa, of conflicting and violent relations) between different agents, both human and non-human. It is about unlearning our species' arrogance by de-centring ourselves from the stories we tell, so as to be able to accept and act on the fact that—as Chakrabarty writes—"humans, all humans," dwell on the planet "more in the position of passing guests than possessive hosts" (2014, 23).

II. An Overview of the Volume

This volume assembles chapters by established experts as well as promising early-career scholars coming from western and central European universities. It is based on the interactions between various and yet complementary traditions of animal studies (French, Anglo-Saxon, and Central European) and combines an innovative reading of canonical works with "minor corpora," giving priority to texts and contexts that have been so far neglected or never interpreted through the lens of human-animal studies. The time frame investigated—from the nineteenth century to the contemporary era—coincides with major transformations both in society (massive industrialization and urbanization in European Countries; the progressive withdrawal of the wilderness; the growing practice of pet-keeping; the birth, success, and evolution of zoos and aquariums; the explosion of meat consumption; the introduction of legislation on animal welfare; the recent concerns and anxieties about the current biotic crisis) and in scientific or philosophical thought (the Darwinian and Freudian revolutions, as well as the new insights into animals' minds and behaviour offered by burgeoning disciplines such as ethology and comparative psychology, along with other theoretical shifts). Such changes have radically reshaped our approach (concrete, philosophical, artistic, and literary) to non-human animals.

The chapters that compose the book draw from a wide range of disciplinary approaches (history and micro-history, visual studies, film studies, postcolonial studies, literature, philosophy, trauma studies, game studies) to show that there exist multiple and complementary ways to overcome anthropocentrism in (and through) the so-called humanities. As mentioned earlier, the chapters revolve around three macro-themes or, it would be better to say, three different and yet closely related strategies of (de)familiarization. It goes without saying that the monograph does not have the ambition to be comprehensive and all-encompassing, but rather broad and suggestive, so as to stimulate further research that follows and complements the paths proposed here.

1. The first section, entitled "'Playing a Part in the Dramaturgy of the Real': (Hi)story on the side of Animals," challenges the role traditionally assigned to animals in historiography, scientific discoveries, political economy, and literary criticism, while mobilizing a variety of historical and scientific documents, as well as the resources of the imagination, in order to make animal lives visible, thinkable, and speakable. Just as "the life of infamous men" sketched by Foucault was "destined to pass beneath all discourse and to disappear without ever being spoken" if the dispositives of power would not have illuminated them at least for a moment, so the lives of animals emerge from the obscurity of the past and can leave behind "traces—brief, incisive, and often enigmatic—at the point of their instantaneous contact with power" (Foucault 2006, 79–80). Hence the importance given in this section to the various institutions and discourses (historically documented or literarily staged) in which animals are enclosed, gazed upon, and exploited or simply removed as referents. In *menageries* and zoos, animals are displayed as evidence of the West's conquest of the "barbaric world"; in laboratories, they are treated as inanimate objects at humans' disposal, while in hyper-industrialized societies in general, they are completely de-animalized; that is, they are either radically humanized as pets or wiped away as processed meat.

If we move to the discursive formations and the disciplinary habitus, we cannot fail to notice that historians have been long reluctant to fully acknowledge the role of non-human actors in shaping human history, and even more reluctant to consider animals "as active, world-producing beings" (Fudge 2017, 260), and therefore as singular individuals having an history *per se* that deserves to be brought to light. In reports and accounts of scientific breakthroughs, the violence against them is almost univocally justified in the name of progress; in literary criticism, animals as characters are frequently re-coded within a symbolic order that transcends them and points towards human beings; and in the political economy, they are simply dismissed or reduced to instruments, if not to raw materials.

The chapters included in this part move away from these human-centric visions of history towards a kind of (post-)humanities that would be able to take into account *all* the historical actors, regardless of their degree of consciousness and awareness. The anti-anthropocentric exercises proposed by the authors consist both in analysing (and questioning) animals' position within the human structures of power that inevitably determine, shape, and condition animal agency,[5] and in making the effort to de-centre (hi)story (in the sense of both human history and narrative plot). More specifically, they try to investigate the unexplored margins of human history by encouraging readers and scholars in human sciences "to counter-focalize" (Spivak 2002)—that is, to imagine or re-read historical events, literary texts, and scientific findings not from humans' perspective but from animals'.

Éric Baratay opens the volume with a chapter entitled "A Giraffe's Journey to France (1826–1827): Recording the Encounter From the Animal's Point of View," which introduces the three main issues of the monograph. In the autumn of 1826, the giraffe offered to Charles X by the Egyptian Pasha arrived in Marseilles, where it spent the winter acclimatizing itself to the population. During the spring of 1827, the giraffe toured many cities, including Aix and Lyon, on its way to Paris. This giraffe—the first living specimen ever seen in Europe—sparked a great excitement in France, before progressively vanishing from the sources once the curiosity faded. This unprecedented encounter—which was simultaneously a confrontation between two worlds, the wilderness and the Western world—has always been told as if humans were the only protagonists. What if we tried to reframe the story and to reconstitute a giraffe-centric account of the event? This is precisely the challenge that Baratay took up in his chapter. Of course, Baratay cannot but rely on traces left by humans in order to even access "the animal side," and yet, by reading the historical sources against the grain, and by interspersing them with current ethological studies, the author composes a biographical narrative that, relying on the method of thick description, succeeds in highlighting both the giraffe's point of view (with its fears, misunderstandings, passions, and resistances) and mutual misunderstandings, tensions, surprises, and adjustments that occurred between this individual animal and the humans who took part in the event.

A similar interdisciplinary work with different sources (quantitative studies of animal longevities, ethological insights, microhistorical and biographical details) characterizes Violette Pouillard's chapter "Structures of Captivity and Animal Agency: The London Zoo, ca. 1865 to the Present Times," the primary aim of which is to restore a balance in zoological garden studies, where animals paradoxically are almost absent. Indeed, despite the opportunities offered by historical sources, academic fieldwork has given priority to a vision of animals as clues to the understanding of cultural, scientific, social, and political issues. In line with the recent development of historical works paying attention to animals *in themselves*, this chapter, like Baratay's, tries to decentre human concerns in order to shed light on animals' lives by analysing a particular case study, that is, the history of captive primates (whose taxonomic proximity to humans has encouraged intimate relationships with both zoo keepers and visitors) at the London zoo during the twentieth century. However, if Baratay lets the animal life emerge through the hypothetical yet accurate and trustworthy biographical account of the giraffe, Pouillard does so by addressing the materiality of primates' lives in all its complexity, that is, by focusing on the multiple devices of captivity and their transformations along the century (architectural structures, spatial arrangements, hygienist measures) insofar as they have probably affected animals' lives more extensively than any other contingent interaction with humans. This approach also helps define the concept of animal agency and unveils the impossibility of separating the dominance over other beings—even when infused with affection and paternalism—from its effects on animals' bodies and minds.

The following chapter, "Atlantic History From the Saddle: The Role of Horses in the Slave-Trading Atlantic World" by Rachael L. Pasierowska, engages in another exercise of (de)familiarization in the field of historical studies. While scholars have long studied the movement of peoples and slave-produced crops across the Atlantic World, there remains another forced migrant that continues to receive little or indeed no direct attention: animals and, in particular, horses, who are used in this chapter as a lens to re-analyze the Transatlantic slave trade, as well as the historical triangulation among Europe, Africa, and the Americas more generally, demonstrating the extent to which—to borrow the words of Sandra Swart—"History was made with horse power; and equally the horses were shaped by human history" (2010, 37). Employing a wide range of secondary and primary sources such as travel accounts, slave interviews, folklore, and visual materials, this chapter purports to foreground how, across the Atlantic World, different peoples and groups (slaves and masters, colonizers and colonized people, Native American peoples, and European "conquistadors") developed similar perceptions, beliefs, practices, and values regarding horses. The three patterns identified by the author (horses as agents of power; horses as a disputed symbol of power and prestige; and, finally, horses as agents *and* symbols of the magical and spiritual worlds) allow a more inclusive historical narrative of the Atlantic world that tries to take into account all actors and newly appraise the contacts and interactions between Europeans, Native American peoples, African societies, and slaves.

In line with this effort of making animals visible and thinkable, Michał Krzykawski's chapter, "Re-animalizing Animals, Re-animating Humans," aims to consider animals as objects (and subjects) of a new critique of political economy, with a special emphasis on the industrial process of killing animals, in what Stiegler defines as the "hyper-industrial epoch," where the invisibility of "processed animals" stands alongside "a selected anthropomorphism which makes us *exclusively* look at our pets."

Krzykawski's reasoning follows Florence Burgat's effort to develop conceptual tools to help us to understand why humanity is carnivorous. However, unlike Burgat, whose attempt is to challenge "the carnivorous fact" in relation to "the initial fiction" that established "humanity as a whole" and resulted in "metaphysical and moral implications soon accompanied by political power," the author rather focuses on the human-global carnivorousness as a disruptive innovation of the capitalism of the second half of the twentieth century. If our carnivorousness is an anthropological fact, as Burgat convincingly puts it, the industrial epoch has totally changed our ways and possibilities of (not) seeing animals. Through a cross-reading of Karl Marx's "general intellect" and Éric Baratay's "animals-machines," Krzykawski tries to show the shift from the reification of animals to their de-animalization by means of the "general intellect," which took place over the course of the twentieth century and made the explosion of meat consumption possible. This detailed analysis of the anthropological machine's functioning in hyper-industrialized societies stems from the conviction that a renewal of the political economy is the condition for ethics, insofar as we cannot expect to act ethically towards animals if we are not even able to see them, that is, to understand the conditions of production and consumption under which they have been de-animalized and made to disappear.

The last two chapters of the first part tackle the entanglement between science, politics, and literature, trying to show to which extent animal lives are co-opted and re-coded into political, scientific, and fictional narratives (or interpretations) that validate animals' invisibility and justify the current structure of species domination.

Anita Jarzyna's chapter, "Laika's Lullabies: Post-anthropocentric Representations of the First Dog in Space," focuses on the story of the homeless dog Laika, the first living creature sent into orbit, in a Soviet satellite called Sputnik 2 in 1957. Laika has immediately become a sort of an icon insofar as her story lends itself well to represent the fate of all animals used for scientific experiments in general, and for space missions in particular: those animals are, in fact, presented as heroes who almost intentionally sacrifice their lives, as is shown in the *Monument of Space Conquerors* in Moscow (1964). At first, the chapter proposes a critical assessment of some recent historical accounts that engage in new and multilateral narratives of the history of the Space Race, insofar as they not only finally acknowledge the crucial role played by animals in the history of spaceflight but also analyze (more or less critically) the prevailing rhetorical strategies employed by official propaganda. In a second step, the chapter moves on to discuss two Polish poems dealing with Laika's story: Zbigniew Herbert's *First the Dog* (1957) and Adam Włodek's *Epitaph* (1968). Both works deliberately exhibit the reassuring rhetoric employed by propaganda and stage the dog's subjectivity. Finally, the chapter compares the two poems to a graphic novel by Nick

It is well known what John Berger wrote about zoological gardens: "the zoo to which people go to meet animals, to observe them, to see them, is, in fact, a monument to the impossibility of such encounters" (1992, 21). And yet—despite the fact that, as Berger noted, the "encounters" are necessarily vitiated by the captivity—there is at least one category of persons that unquestionably has daily and close interactions with the zoo animals: the zoo keepers. Interestingly enough, the biography of these fundamental actors of zoological gardens, through whom the visitors could experience a relation by proxy with the animals, is almost unknown, and briefly recorded only when something went wrong, such as when an animal unexpectedly killed the guardian, causing great consternation but also raising concerns about animal resistance and agency. Maria P. Gindhart's chapter, "Animals and Their Keepers in Belle Époque Postcards of the Jardin des Plantes Menagerie," drawing on postcards in the collection of the National Museum of Natural History, seeks to highlight the bonds between keepers and zoo animals in early-twentieth-century Paris. Multiple postcards feature one or more animals being fed by a keeper. Moreover, although fencing or grilles are frequently visible, there is no barrier in the foreground, thus providing zoogoers with a surrogate proximity to the animals without the actual danger that led some *ménagers* to their death. Postcards in Gindhart's chapter are both analyzed for their visual content and used as a starting point for further investigation. By supplementing them with press reports, or by tracing their re-use in contemporary children's books (such as *Kako le terrible*), the author gives further insights into the biographies and relations of zoo animals and keepers, as well as into the fictional re-appropriations of their history for educational purposes.

Quentin Montagne's chapter, "'Seeing Eye to Eye, Through a Glass Clearly'? Blurring the Boundary Between Humans and Animals," puts the emphasis on an another "encapsulated world" that promises the visitors an intense proximity with animals, while still maintaining a very tangible yet transparent barrier: the aquarium. Despite its growing popularity and continuous improvement since the nineteenth century, the aquarium—officially invented by Robert Warrington in 1850—is still little discussed in the field of animal studies. Regardless of its scale, the aquarium puts us in contact with the exhibited fauna while keeping us away from it. Either more or less visible, this boundary quickly became a major concern of wildlife park designers hoping to minimize its presence. Montagne's chapter analyzes the irreducible tension between the different apparatuses and technical devices aimed at bringing man and fauna closer in public aquariums since the nineteenth century, and the enduring divide that separates human viewers and enclosed animals (while bars or wire fences can be replaced by ditches, the aquarium cannot avoid its glass or, today, acrylic screen). Through numbers of historical and contemporary examples (from *L'Aquarium de Paris* (1900) to the OdySea Aquarium in Scottsdale, Arizona, to mention but a few), this chapter explores the animal-human relations constructed and implied by ancient and modern aquariums and takes into account various innovations seeking to blur the frontiers between humans and animals. Two different strategies seem to emerge. The first, related to home aquariums, consists of an increasingly intense integration of the tanks into the interiors. The second, related to public aquariums, is characterized by the immersion of the visitor into marine life by means of new tanks and spectacular and interactive attractions.

With the third chapter of this part, we move to another institutional context that frames and allows reflection regarding the mutual constitutiveness of human and animal: museums and galleries. To be more precise, Kari Weil's chapter "Flesh, Fur, and Forgetting," in order to address the shared embodiment of humans and other animals, engages in a multilevel analysis of different forms of entanglements, interactions, and interdependencies: between the human visitors and the exhibited bodies, visual art and written texts, living creatures and dead bodies, and memories and forgetting. Kari Weil's chapter strives to dwell precisely on these intervals, on the spaces of indeterminacy between *zoe* and *bios* "that the anthropological machine would prefer to ignore" (Weil 43), and does so through a close reading and analysis of the work of the Belgian artist Berlinde De Bruyckere, paying special attention to her lasting and intense collaboration with J.M. Coetzee, who helped her curate her show at the 2013 Venice Biennale and with whom she co-authored several books. Among them, Weil focuses in particular on *We Are All Flesh*, in which excerpts from Coetzee's many novels are interspersed with images from De Bruyckere's many installations in such a way that the unexpected juxtapositions of images with writing resonate with the purpose of evoking the human-animal secret bond and their shared vulnerability. The first image in *We Are All Flesh* is a close-up from an installation called "The Wound," showing what appears to be a stitched and healing scar, here made of wax, wood, and cloth. Curiously, the book moves from this notion of human scarring to images of the glossy, taxidermied coat of a horse, shot from angles that render their curled or folded positions all but illegible. What is the relation of this fur to the idea of scarring? Do horses not scar and heal in the same way, we might ask? Is this why they are "receptacles of the forgotten," as Walter Benjamin says, their wounds still open but hidden beneath their fur rather than stitched over by a healing and written history? Does this creaturely matter "speak," and how? These are some of the questions this chapter raises in moving between Coetzee, De Bruyckere's sculptures, and nineteenth-century writings about memory, habits, and the training of horses.

Finally, it is through the lens of the literary establishment that the last chapter of this part tackles the issue of how animals (unwillingly) participate to the construction of human identity. The case study proposed by Eva Voldřichová Beránková—"Animals as Monsters, Victims, and Models in the Age of Decadence"—focuses on France *fin de siècle* and demonstrates to which extent Decadents, starting from a tactical reversal of the stigmas attached to them (they are described by critics and rival writers as a strange or disgusting animal species), end up developing their new and provocative poetics precisely around the repugnant fauna to which they were compared, using species differences also to conceptualize social and gender issues. Moreover, the chapter suggests that, at the turn of the century, these "literary disputes" might turn out to have some links to the Decadents' actual involvement in the animal cause. Presented by opponents—or by themselves—as exotic monsters, monkeys, snails, or tapeworms, to name just a few examples, the authors and characters of *fin de siècle* seem to be going through a sort of inverted evolution, from primates to canines, from mammals to molluscs, from frogs to tadpoles. In parallel, animals, omnipresent in decadent literary works, are alternatively objects of self-projection, victims of sophisticated and perverse experiments, weapons directed against the hated human race, or simply protagonists of nightmares. By identifying the most recurrent paradigms concerning animals, Beránková's chapter investigates how French Decadence reflects upon identity and creativity through the lens of repellent fauna and tries to understand what insights into animals' lives and minds we can deduce from the Decadents' predominant metaphors and uncanny rhetorical figures.

3. The third and last section of the monograph, "'Across the Narrow Abyss of Non-comprehension': Entering Animals' Lives," turns to the possibility of exploring, grasping, and portraying animals' lives, minds, and language. At issue is not only the search for interpretative frameworks and representational codes that enhance what is humanly thinkable, perceivable, and palpable but also a process of translation that aims to render animals' experience and perception into human words and visual language, while bearing in mind that this translation process is necessarily open-ended and fraught with traps. Nevertheless, instead of stressing animals' impenetrability as is commonly the case, the chapters included in this part insist, to the contrary, on the cognitive value of literary language, critical anthropomorphism, shamanism, POV shots in films, video games, and animal simulators as multiple ways of leading us *out* of our own worlds and minds and *into* those of other animals. In dealing with the difficulties faced by human language when trying to understand and depict animals' mysterious relationship with their milieu, the chapters tackle the *aporias* of our cognitive experience and endeavour to find a third way between two opposite approaches: *anthropocentric arrogance* (according to which the human mind and language would hold unlimited power, having the capacity to imagine and account for every living being on this planet) and *post-human melancholia* (human knowledge would be radically anthropocentric, bounded, and inadequate, and for this reason,

animal lives will be always unattainable and beyond our understanding, despite our best efforts).

In other words, human language is certainly one of the most powerful tools through which we explain, master, and dominate the world, but it is also the place of a fundamental dispossession: "Je n'ai qu'une langue, ce n'est pas la mienne" ("I have only one language, it is not mine") wrote Derrida in *Le Monolinguisme de l'autre* (1996), meaning by this performative contradiction that our language is not only contingently heteronomous (it could be the language imposed by the oppressor) but it is so in a more fundamental way, since every language we consider as ours comes from "the other," precedes us, and configures in advance the very possibility to say "I," to define ourselves. Why would this be pertinent for animal studies? If we are ready to conceive of human language in this way, then, paradoxically, our experience as beings of language, far from alienating us from the rest of living beings, can make us unlearn species arrogance, scale down our egocentric pride, and acknowledge that the surrounding environment shapes us at least as much as we shape it. Moreover, if we start from the assumption that all terrestrial creatures share a fundamental "arché," that—as Anne Simon reminds us in her chapter drawing on Brian L. Keeley's studies—humans are not only anthropomorphic but also primatomorphic, mammalomorphic, and, more broadly, biomorphic, then why would human language not be able to allude to, or account for, extra-linguistic experiences, abilities, perceptions, and affections?

The first three chapters of this part explore precisely the potential "animal nature" of human language, thematizing the triple challenge that zoopoetics has taken on: how to push human language out of itself, how to re-interpret literary works and literariness in general through the lens of the "animal question," and, finally, how to rethink such anthropomorphic categories as those used in narratological studies (narrative voice, for example) in a way that they can account for and include nonhumans, rendering audible the unheard, visible the invisible, thinkable the unthinkable.

Anne Simon opens the section with a chapter entitled "Literature and Animal Expressiveness: On Cognitive and Ethic Aspects of Zoopoetics," which highlights the plurality of stylistic, linguistic, and narrative tools that writers employ to express and stage the plurality of animal behaviours, affects, and worlds. It also purports to show that if writers and readers can be engaged in animals' worlds, this is precisely because animals are not mere *aloga*, but they are also "figurative and rhetorical" beings: they display communication strategies, for example, passing on their patterns from generation to generation, leaving traces that "tell stories." Perspectivism, metamorphosis, and hybridity are not simply literary or mythological themes and devices but rather universal patterns and experiences shared by all living beings. Moreover, if we insist on the paradigm of "the indistinction of human and animal life prior to their separation," then we should also admit that humans and non-humans share a wide range of emotions, expressions, and affects that figurative language might be able to account for. In this respect, Anne Simon invites us to substantially reassess the long-standing issue of anthropomorphism, which we would never be able to elude. If we manage to use it with caution, anthropomorphism, far from leading necessarily to misconceptions and misrepresentations, might turn out to be productive in both scientific and literary studies, especially if we consider humans beings and their abilities, including language, in terms of co-evolution and symbiosis with other living beings. This is not to erase differences but to acknowledge that all life forms "are indeed in a relationship of dependence with an *archè*—an origin, a reason, a refuge, a dwelling, the Earth." "Thus, Zoopoetics—concludes Anne Simon—proves to be also a *zoopoethics*, insofar as it leads readers to stop looking at their habitat—their *oikos*—from above, as if they were not part and parcel of it."

Among the many literary strategies and categories evoked by Anne Simon, Kári Driscoll's chapter, "'Une langue ou une musique inouïe, assez inhumaine . . . ': Narrative Voice and the Question of the Animal," focuses on the issue of narrative voice, crossing narratology, studies on mice in neuroscience, and French philosophy in unexpected ways. In *The Animal That Therefore I Am*, Derrida wonders whether it would be possible to think of the discourse on the animal in musical terms, and if so, whether one could change the key, or the tone of the music, by inserting a "flat" (*bémol*). The task would be to render audible "an unheard language or music" that would be "somewhat inhuman" but a language nonetheless, "whose words, concepts, singing, and accent can finally manage to be foreign enough to everything that, in all human languages, will have harbored so many *bêtises* concerning the so-called animal" (Derrida 2008, 63). Driscoll's chapter pursues this intriguing proposition by means of a reading of Franz Kafka's final story, "Josefine, the Singer or the Mouse Folk," paying careful attention to the controversy regarding the status of Josefine's vocalizations (Is it really singing? or perhaps rather squeaking, or whistling?), which, moreover, is mirrored in the scientific discourse surrounding the songs of actual mice. What is at stake in rendering this inhuman music audible, and how can we conceive of this procedure as a zoopoetic enterprise? The voice plays a crucial role in almost all of Kafka's animal narratives, in part because of the ancient tension between speech (*logos*), which is considered unique to humans, and voice (*phonē*), which we share with other living beings. Hence, the status of Josefine's song, and of mouse song in general, obeys the logic of the anthropological machine. The task of this zoopoetic reading, then, is to show how Kafka's text unsettles, or might render inoperative, that logic, and how to disjoin the unquestioned link between voice and human subjectivity by learning to listen *otherwise*.

Jonathan Pollock's chapter, "From Being a Beast to Becoming-Animal: Literary Experiments in Crossing the Species Divide," looks more closely into one of the issues raised by Anne Simon's chapter, to wit, the allegedly impossibility to get rid of anthropocentrism and anthropomorphism in our efforts of depicting and narrating animals. Pollock's chapter contests this stance by identifying the pitfalls into which Charles Foster has fallen when criticizing the "literary shamanism" of J.A. Baker in his classic work *The Peregrine* (1967), the tale of how the author obsessively pursues some falcons until he suggests he has become one. Charles Foster, in his study *Being a Beast* (2016), which is the account of his own efforts to become five different animals, attempts to identify with certain species of animals by concretely adopting their lifestyles and habitats. He lives in a sett as a badger, swims in rivers as an otter, attempts to fly as a swift, and forages in bins as a city fox—ostensibly to find out what being a wild animal is really like. To his mind, such attempts are doomed to failure from the outset. However, Pollock demonstrates that Foster's rejection of "literary shamanism" is based on some presuppositions or misconceptions that are also more generally found in the endless debate about humans' impossibility to avoid anthropocentrism and anthropomorphism. These misconceptions concern Foster's understanding of Uexküll's animal worlds, his conception of becoming, his view on shamanism, and finally his attitude towards the practice of writing. Deleuze and Guattari have already warned us that becoming-animal had little to do with imitating animals (the trap Foster seems to fall into). More recently, anthropologists such as Eduardo Viveiros de Castro and Philippe Descola have explored the philosophical implications of shamanism, characterizing indigenous American theories as "perspectivist" (rather than essentialist) and "multi-natural" (as opposed to multi-cultural). In this respect, they point out, the multiple natures of animist ontology are reminiscent of the multiplicity of animal worlds in ethological science. Shamanism, then, understood properly and despite Foster's claims to the contrary, might well provide us with the conceptual tools by which to assess the singular experiences of those who attempt to cross the species divide in order to adopt other-than-human perspectives.

With the last two chapters, we leave the literary field to explore other languages and creative devices that would allow us to make the imaginary experience of seeing and perceiving the world through animals' eyes, bodies, and minds: cinema and video games.

Concepción Cortés Zulueta's chapter, "Cameras That Pose as Animals: Imagining Non-human Animals Through the POV Shot," looks at one particular cinematographic strategy, the POV shot embodying animals' gaze, and reflects upon how images and narratives oscillate between our symbolic projections on animals and their perceptions or biological realities. By means of several examples—ranging from early British scientific documentaries to more recent movies, such as Almodovar's *Qué he hecho yo para merecer esto*, Herzog's *Bad Lieutenant*, or *The Simpsons* episode "Bart's Dog Gets an F"—Cortés Zulueta identifies the main functions that can be assigned to the animal POV shot: the animal as an actual threat to humans' lives (as it is shown in horror films), the animal as a full gazing-subject whose unfamiliar sensory system challenges human certainties and perceptions of reality, and finally the animal as a "careless witness", whose point of view serves to decentre humans from their own narratives. Cortés Zulueta's chapter aims to analyze the animal POV shot in all its complexity. First, the author sketches a brief history of the ways in which the animal POV shot has been constructed and codified, highlighting some recurrent patterns; second, she discusses the foundational role of the POV shot in film editing and the ways in which it has partially undermined the human basis of the filmic language itself, obliging directors to find new cinematographic solutions to the challenges posed by filmed and gazing animals; third, she insists on the influences that these cinematic representations (more or less accurate and trustworthy) have had on our current assumptions about animals' perceptions. Moreover, the author considers the cognitive value of the POV shot with respect to animals: on the one hand, the POV shot could not be anything but a (mis)representation; on the other hand, animal subjective shots in movies might help deepen our comprehension of animals.

The monograph concludes with the chapter by Michael Fuchs, "Playing (With) the Non-human: The Animal Avatar in *Bear Simulator.*" The author asks whether video games featuring animal avatars may open spaces of affective alliances with more-than-human others (both animals and machines) and thus challenge our traditional understanding of human-animal encounters. The idea that video games are potential vehicles for "becoming-with-many" is exemplified by discussing *Bear Simulator* (2016), a video game that promises players that they will experience life through the eyes of a bear, "controlling" a bear roaming a forest. By insisting on the ambivalence of game playing (players "master" their avatars at least as much as the latter "master" the players), and by combining Gilles Deleuze and Félix Guattari's notion of becoming-animal with phenomenological approaches to video game avatars and object-oriented ontology, Michael Fuchs suggests that *Bear Simulator* simultaneously expands and collapses the limits of the human body, allowing players to playfully explore the fusion of the human and non-human. This, however, would be too optimistic a conclusion. It is only by taking into account both the negative reactions to the first version of the game and the modifications introduced by the designers that the human resistance against the hypothesis of upsetting the established anthropocentric worldview—an important leitmotif of this monograph—emerges in all its strength. Indeed, by focusing on exploring and nurturing the relationship between the human player and the animal avatar, *Bear Simulator* rejects the fetishization of objectives and the telos obsession typical of video games. However, as shown by the numerous criticisms reproaching the game's absence of plot, the encounter with non-human others proves to be a very uncomfortable (if not unbearable) experience for humans.

Despite the difficulties and the understandable resistances, however, the authors of this monograph firmly believe that risking ourselves in the territory of non-human animals is not only necessary—more than ever before—but also quite an exciting adventure.

Acknowledgements

This volume was supported by the European Regional Development Fund-Project "Creativity and Adaptability as Conditions of the Success of Europe in an Interrelated World" (No. CZ.02.1.01/0.0/0.0/16_019/0000734), by CEFRES—French Research Center in Humanities and Social Sciences in Prague (USR 3138 CNRS-MEAE), and by the research project PROGRES Q14 "Crisis of Rationality in Modern Thought."

Notes

1. On the absence of animals in Agamben's reflection about the anthropological machine, see Oliver (2007); Rattasepp (2014); Calarco (2015, 53–5).
2. Bailly (2013), Kindle position 79/1569: "les animaux, c'est presque leur définition, n'ont pas la parole …, il ne s'agit d'ailleurs pas, en s'approchant d'eux, de la leur donner …, il s'agit d'aller au-devant de leur silence et de tenter d'identifier ce qui s'y dit."
3. Bailly (2013), kindle position 861/1569: "ce serait amusant, mais parce qu'à l'issue de chacun de ces chemins, notre vision du paysage ressort élargie, enrichie, émancipée."
4. Interview to Laura Pugno in Alessandro Macilenti (2018, 183): "Non possiamo più godere di un antropocentrismo tranquillo." [Translation mine].
5. On the issue of animal agency, see McFarland and Hediger (2009) and the insightful article by Carter and Charles (2013).
6. For a wide range of different typologies of human-animal encounters, see Tyler and Rossini (2009).

References

Agamben, Giorgio (2004). *The Open: Man and Animal.* Stanford: Stanford University Press.

Bailly, Jean-Christophe (2013). *La Parti pris des animaux.* Paris: Christian Bourgois Éditeur.

Berger, John (1992). "Why Look at Animals?" In *About Looking*, 3–28. New York: Vintage Book.

Burt, Jonathan (2005). "John Berger's *Why Look at Animals*: A Close Reading." *Worldviews*, no. 2: 203–18.

Calarco, Matthew (2015). *Thinking Through Animals: Identity, Difference, Indistinction.* Stanford: Stanford University Press.

Carter, Bob, and Charles, Nickie (2013). "Animals, Agency and Resistance." *Journal for the Theory of Social Behaviour*, no. 3: 322–40.

Chakrabarty, Dipesh (2014). "Climate and Capital: On Conjoined Histories." *Critical Inquiry*, no. 1: 1–23.

Derrida, Jacques (1995). "'Eating Well', or the Calculation of the Subject." Translated by Peter Connor and Avital Ronell. In *Points . . . Interviews, 1974–1994*, edited by Elisabeth Weber, 255–87. Stanford: Stanford University Press.
Derrida, Jacques (1996). *Le Monolinguisme de l'autre*. Paris: Éditions Galilée.
Derrida, Jacques (2008). *The Animal That Therefore I Am*. Edited by Marie-Louise Mallet. Translated by David Wills. New York: Fordham University Press.
Domanska, Ewa (2011). "Beyond Anthropocentrism in Historical Studies." *Historein*, no. 10: 118–30.
Foucault, Michel (2006). "The Life of Infamous Men." In *Michel Foucault: Power, Truth, Strategy*, edited by Meaghan Morris and Paul Patton, 77–91. Sideny: Feral Publications.
Fudge, Erica (2017). "What Was It Like to Be a Cow? History and Animal Studies." In *The Oxford Handbook of Animal Studies*, edited by Linda Kalof, 258–78. Oxford: Oxford University Press.
Haraway, Donna (2008). *When Species Meet*. Minneapolis/London: University of Minnesota Press.
Macilenti, Alessandro (2018). *Characterizing the Anthropocene: Ecological Degradation in Italian Twenty-First Century Writing*. Berlin: Peter Lang.
McFarland, Sarah, and Hediger, Ryan (eds.) (2009). *Animals and Agency: an Interdisciplinary Exploration*. Leiden/Boston: Brill.
Morton, Timothy (2018). *Being Ecological*. London: Pelican.
Oliver, Kelly (2007). "Stopping the Anthropological Machine: Agamben with Heidegger and Merleau-Ponty." *PhaenEx*, no. 2: 1–23.
Rattasepp, Silver (2014). "The Anthropological Machine and the Absence of Animals." In *Non-Humans in Social Science: Ontologies, Theories and Case Studies*, edited by Karolína Pauknerová, Marco Stella, Petr Gibas, et al., 29–45. Červený Kostelec: Pavel Mervart.
Shklovsky, Viktor (2017). "Art as Device." In *Victor Shklovsky: A Reader*, edited and translated by Alexandra Berlina, 74–96. New York: Bloomsbury.
Spivak, Gayatri C. (2002). "Ethics and Politics in Tagore, Coetzee, and Certain Scenes of Teaching." *Diacritics*, no. 3–4: 17–31.
Swart, Sandra (2010). *Riding High: Horses, Humans and History in South Africa*. Johannesburg: Wits University Press.
Tyler, Tom, and Rossini, Manuela (eds.) (2009). *Animal Encounters*. Leiden/Boston: Brill.
Weil, Kari (2012). *Thinking Animals: Why Animal Studies Now?* New York: Columbia University Press.

Part I

"Playing a Part in the Dramaturgy of the Real"

(Hi)story on the Side of Animals

1 A Giraffe's Journey to France (1826–1827)

Recording the Encounter From the Animal's Point of View[1]

Éric Baratay

In 1826, Charles X received a giraffe as a diplomatic gift from the Egyptian Pasha. It arrived in Marseilles that October, where it spent the winter to get acclimatized. It was taken for walks every afternoon among crowds of people, and in the evenings, social events were organized in its honour. The giraffe was sent to Paris on 20th May 1827, along with naturalist Etienne Geoffroy Saint-Hilaire, mahouts, police officers, and management carts. Crowds rushed along the way, and it was given a triumphant welcome which was long-remembered, so vivid was the curiosity, imagination, and passion aroused by the animal. The press published accounts of the trip, and the King was informed on a regular basis. The giraffe entered Paris on June 30 and was presented to the King at Saint Cloud on July 9. By December, the Jardin des Plantes had welcomed 600,000 visitors from across the country, rushing to witness the giraffe's afternoon stroll. The excitement was triggered by the event itself, since it was the first living giraffe seen in the Western world probably since the Roman Empire. It stirred wild enthusiasm: its effigy festooned almanacs, calendars, and prints that were sold around the country. The giraffe even became the star of plays, of satirical pamphlets (*La Giraffe ou le gouvernement des bêtes*, 1872), of comic booklets (*Dame giraffe à Paris*, 1827), and of songs and tunes (*La Giraffe*, waltz for piano). Crockery adorned with fanciful portraits was mass-produced, while women's fashion went wild with giraffe-like colours, sleeves, garnitures, necklaces, ribbons, sunshades, and hairstyles!

Of this adventure, history mainly retained such aspects—omnipresent in documents—pertaining to the human perspective, focusing on the relationship between France and Egypt, on the political significance of the gift, on the diligent actions of authorities, on the individuals' twists and turns in the journey, on the public's reaction and on the fashion trend it caused (Allin 1999; Dardaud 2007). This version is also adopted in youth literature and cinema and by naturalists who never thought of a different approach to look at events (Bezançon 2012; Hartenberger 2010). It pertains to the usual narrative about exhibitions depicting exotic beasts (Cowie 2014; Robbins 2002): indeed, the history of animals which has developed over the last 30 years in the Western world is actually a *human* history of animals, in which the latter are overlooked as real beings.

Still, the giraffe's story is not only human. The general public did not come to see a stuffed object but a living being, whose actions and reactions caused a stir. It was, first and foremost, an encounter between an animal and a population that belonged to two different worlds. In their mutual discovery that led to reciprocal surprise and reactions, each of them created a relationship that needs to be apprehended entirely: each side needs to be accounted for. It is thus essential to look at events from the giraffe's perspective in order to better understand the situation. The personal history of a given actor has become a legitimate subject of study in the humanities and social sciences, as it gives way to a better understanding of its experience and of its role in an event, a phenomenon, and an era. Learning about the actor's reactions enables us to envisage their influences on the surrounding environment. This method must be applied to animals, in general, and to this giraffe, in particular.

Embracing the animals' viewpoint means trying to stand alongside them to adopt their geographic perspective and understand what they are experiencing, are subjected to, and how they act and react. It also means trying to project ourselves into them in order to discover their psychological perspective, what they see and what they feel. It is but a desire, an effort at projection, a method—similar to that of naturalists, hunters, taxidermists, and ethologists—that may help us shift our focus and can deeply contribute to our research. I have written elsewhere (Baratay 2012, 2013, 2017) on the epistemological and methodological aspects of this topic, assessing at the same time recent reflections on animal actors and their relationships with humans.

1. The Question of the Sources

A substantial amount of information survives thanks to the testimonies left by civil servants, such as the Bouches-du-Rhône prefect, and by two naturalists: Salze, a teacher at the school of medicine in Marseilles, wrote a memorandum at the request of the Parisian museum, whereas Étienne Geoffroy Saint-Hilaire, a professor at this museum and one of the most renowned naturalists of his time, along with Cuvier, had been requested to pick up this exceptional animal. He wrote letters during the trip and published an analysis of the process after arriving in Paris (Salze 1827; Geoffroy Saint-Hilaire 1827).[2] Using such documents to rediscover animal actions may seem paradoxical regarding their reliability; their incomplete, limited, and biased scope calls them into question. Humans are only interested in certain species, races, and individuals, and then only in a few aspects on which they hardly recorded all the facts, only keeping in mind what they could and were willing to see and distorting with their imagination and biases the certainties they had about a species, a society, and an era. But these problems also emerge with respect to human history, a domain in which historians often have to use intermediary sources: most documents on farmers have been drawn up by dignitaries, yet they are used to write rural history. The difficulty here is greater but not of a different nature: we must do our best with the sources we have.

All witnesses focused first and foremost on the anatomy, the movements, and the diet of the giraffe (Grognier 1827), but they also considered its behaviour and reactions. They did so through human lenses, of course, with the interests, oversights, language, and writing style of a particular era and a particular scientific field. Yet, they were not as narrow-minded as one might fear: the novelty of the giraffe sparked in them an unusual curiosity, while the lack of previous studies gave them a freedom of mind and allowed them to see the animal through a fresh pair of eyes.

They obviously only wrote what they wanted to say, only a part of what they saw or what they thought they had seen—which is only a part of what happened. Nonetheless, these documents give us a glimpse of the unfolding of events—provided they are inverted to illustrate the animal's perspective, that is, to look for evidence that can be inferred by a process that Carlo Ginzburg has called the "evidential paradigm" (Ginzburg 2012). This means going beyond—and not doing away with—the cultural approach. The latter often confines us to deconstruct discourses and focuses exclusively on representations considered as the sole observable phenomena. This undertaking is necessary, but the triumph of cultural readings has often turned this awareness of the constructed character of our knowledge into the main purpose of historical studies. It is thus necessary to grasp reality anew, relying on the concept of "situated knowledge" (Haraway 1991), which allows the construction of a knowledge while being aware of the context of its elaboration.

We need to address ethological knowledge on giraffids, which would allow us to research, comment upon, explain, complete, and suggest revisions to existing studies. Nevertheless, our aim is not to corroborate the documentary evidence with today's knowledge in ethology, as this could easily lead to the rejection or the deformation of the former according to the particular perspective of an era or a school of thought. It is crucial to intertwine situated gazes and knowledge: that of the giraffe's contemporaries, who could have overlooked some aspects or, on the contrary, could have seen what cannot or will not be observed nowadays; and that of current naturalists, who may have greater and better knowledge than earlier ones but who might also neglect some aspects, thus having their limited knowledge distorted by philosophical, theological, sociological, and political considerations, like historians for instance. Another reason why one should not model history on ethology is that it could create a timeless portrait of animals based on their current representatives, thus denying their historicity, while this dimension should be at the core of the yet-to-be-constructed *animal* history of animals. This is especially true since ethologists have recently become aware of this matter and are calling for a social and cultural history of animals (McGrew 2010; Whiten 2010). History and ethology must inspire, influence, and affect one another, in order to develop the animal side of history and to adopt the historical dimension of ethology. As a matter of fact, historians are constrained not only by the state of documentary sources but also by the development of research in ethology. Research on giraffids is still scarce and is only led by a few academics. Ann Dagg's work (2014), now an international reference-point, remains the essential source of this study. To avoid multiple references in this limited format, I note here that most of the ethological indications used here are taken from her book. References from other works will be indicated as such.

2. The Arrival in Marseilles

It was of course but one stage in a long journey for the giraffe. Captured in Sudan by local hunters, our giraffe was walked to the Nile before arriving in Alexandria by boat. It then crossed the Mediterranean Sea in the bilge of a ship, with its head popping out through the deck through a hole especially made for this occasion. This giraffe did survive, while two fellow specimens, sent at the same time to Austria and England, quickly died. Its survival is even more remarkable given, as it is now known, the very high mortality rate of fragile giraffe calves after capture.

We can try to establish the reasons for its endurance. According to documents, it was captured between the age of two and five months, which means it was young enough to find surrogate mothers (in healers, camels, and horses) and yet old enough to endure their constant replacement, for it had already learnt how to be independent with other young animals while the mothers went grazing. The specimen was a female, bound to live in a group and thus more apt to promiscuity than males, who peregrinate apart. These groups tend to regroup in the wilderness—and such was its experience during the journey, from the moment when its captors sold it to the governor, who sent it to the Pasha, who offered it to the French Consulate, who himself left it with a ship-owner! But its peculiar and exceptional disposition, which was noticed by the Arabic mahouts who were used to comparing catches, was responsible for its survival of the hardships—a very important feature that we must keep in mind. Calm and docile, it did not try to run away after being caught, contrary to others. Flexible and obliging, it got used to walking freely with its captors. While "walking, it would never show any desire to flee; but it would often show cheerfulness like a young horse," and thus would rear up, jump, and express a need to exert itself, which overcame the stress and depression that caused many others to die off (Ferlus 1827a, 4; Salze 1827, 71–2; Geoffroy Saint-Hilaire 1827, 210–11, 223).

This means that it had a certain disposition, or personality, using the current vocabulary of ethologists. To look at events from the giraffe's perspective, it is essential to consider, as a starting point, the fact that it was not merely an actor influencing humans and that it did not only have an agency to engage in arrangements and interactions with them, as Anglo-Saxon studies currently maintain (*History and Theory*, 2013). Such notions invite analyses regarding animals' influence upon humans' actions but not the actions of the animals themselves. This giraffe needs to be appraised as an individual with its own features, as a "person" with its own behaviours, and even as a subject making choices and planning strategies. I am not arguing for a conflation of beings but rather suggesting we open these notions (subject, individuality) beyond the restrictions set on them between the eighteenth and nineteenth centuries, in order to instead appreciate the diversity of all, and the individuality of each being.

The historical encounter of the giraffe with the European population starts at the end of its quarantine on 14th November 1826, under the cover of dark, which was meant to prevent a public panic, for fear of a rumour about the arrival of a gigantic monster. The giraffe came out "unafraid," that is, without resisting, deterred by the tension of the tight lead ropes held by the mahouts, and probably driven by a desire to move around—which I will address in the following. It walked all the more steadily that it could not really see this new, possibly upsetting, landscape. It was surrounded by only a few men to avoid scaring it, and it followed behind a horse ridden by a notable of Marseilles, who wanted to officially introduce the animal to the city. The giraffe seemed reassured by the horse—a species it was acquainted with since its capture.

The men of the escort understood this when, at the city gate, the giraffe "stopped abruptly, not wanting to move forward nor turn back: it showed fear, mingled with anxiety." As for many animals, this dark corridor was frightening: the giraffe refused to step into it, stooped on its knees, and resisted as it was repeatedly pulled forward and backward to make it turn around in desperation. It then noticed the returning horse, which had been pushed forward beyond sight to make way, and whose absence upset the giraffe. The rider had taken the horse back to find out what was happening and suggested to use it as a leader, much like those used for herd animals such as sheep, horses, and cows. As soon as "the Giraffe saw the horse again . . ., it was peaceful." It followed the horse "very closely" into the city and thus arrived at the prefect's mansion. This incident demonstrates how the animals which the giraffe had mingled with since its capture were necessary for its survival and adaptation. They constituted a link to its initial experience in the savannah, where the unstable herd instinct of giraffe groups and the encounter with other species from zebras to antelopes allowed giraffes—especially females—to accept, even to seek, the company of other animals. This allowed the giraffe to survive the six-month wait it endured in an annexe of the Prefecture that had been converted into a stable for the occasion: it sheltered three cows, which helped feed it, two antelopes, also destined to the national museum, and two draft horses, which were used to warm up the room (Salze 1827, 68–9, 82–3)![3]

Still, the giraffe was never completely at ease, for it refrained from its natural body-position for deep sleep: lying on its stomach, its head thrown over its right hind leg—observers would have noticed if it had taken this position, even though the species adopts it in the wild only for a few minutes per night to avoid exposure to predators. Yet, this adaptation is nowadays considered as a sign of distinct stress. The giraffe "barely lies down" and thus slumbers while standing—for short periods of time—remaining vigilant while employing the remarkable ability of its species to barely sleep, a more important feature in older specimens than in younger ones, which are supposed to get more sleep. During the day, it seemed to suffer from the narrowness of the room—both in terms of height and surface—since men always put the beasts in places proportional to what was available and not proportional to the animals' needs, which they poorly understood. As a matter of fact, as it was 11′5″ (3.53 metres), it did not stand far from the ceiling (13′/4 metres high) and could not really come and go as it pleased. It would therefore be "almost constantly on the move": it would prance about to lighten the overload of its haunch due to the height of its neck, and which was very tiring when inactive. It moved forward and backward, probably out of boredom, which provoked this typical behaviour (Salze 1827, 80–3; Bashaw 2001). It was thus a peculiar giraffe with an abnormal behaviour that these Europeans observed, the same way their own excitement was quite unusual.

3. Fears, Misunderstandings, Adjustments

While inactive, the giraffe would take an interest in the visitors, who gathered every day in front of the premises. It would often lower its head to look at them, or poke it out of the window, and "would happen to lick strangers from time to time and gladly smell those who came near it." Yet it would withdraw when frightened, hearing shouts or perceiving hasty movements from individuals who were scared by its behaviour. Even though it was based on a mutual curiosity, the encounter between these two worlds only triggered a partial and short-lived overlap and incomprehension. It also gave way to mutual adaptations, as the first walks around the Prefecture's courtyard suggest: the animal would often start galloping to stretch its limbs, taking the mahouts who supervised it along, but would quickly get tired of dragging them around with ropes—especially since giraffes feel discomfort when running because of their unbalanced bodies and their long necks. Henceforth, it would "raise very high and drop, motionless, on its legs," in order to make up for not being able to gallop. It would jump up and down or between two steps with the men that let it do so, before slowly adopting a step-by-step, consensual rhythm with them (Salze 1827, 78, 81–2).[4]

In the same vein, when the animal walked in the streets of Marseille—and from March 1827 on, outside the city—it felt constrained by the ropes held by the six men who served as its escort: they prevented it from running riot and controlled its every move. The giraffe felt less and less stress, as it showed a relative placidity—allowing it to quickly take up new habits—and because it strove to follow the dairy cows that the men (having learnt the lesson of the initial entrance into the city) put ahead of it. The giraffe would "obey their direction" and thus feel a strong kinship with these fellow bovid, placid ruminants, like the antelopes it had mixed with in the savannah and the two specimens in its stable. It was interested not only in these two particular cows but in all the specimens met along the way: they did not show any sign of fear, contrary to the horses or mules, whose straight ears betrayed anxiety. The latter pawed at the ground when having to wait and made a break every chance they got—these equine animals being very sensitive to the giraffes' smell—whereas the giraffe would carefully observe and smell them, would try to follow them, and would watch them go away (Grognier 14 juin 1827).[5] Its anxious curiosity was not only directed towards man!

The giraffe was mostly keen on trees: it would examine each of them, pull on the ropes to stop in front of them, and force the men to let it smell and taste the leaves, something they were actually inclined to do since they took interest in its natural behaviours, one of the few they could observe: the two sides here could find common ground. It thus engaged in an activity that keeps wild giraffes (mostly females) busy—slow chewers as they are—but which it was deprived of in captivity. It would indeed express its frustrations—in the remaining time—through a typical behaviour which has been thoroughly studied by ethologists. Many zoo animals display the same pattern, unable to develop their behaviours in their limited environments: it would lick the walls of the stable or wrap its long tongue around the hands of passers-by during walks—the latter did not understand and believed this to be its usual behaviour.

The giraffe actually enjoyed human beings in certain situations. It "would willingly smell" the closest onlookers it met on the streets and paths and obligingly let come forward as they hurried, fussed, talked, shouted, and were astonished, ecstatic, and afraid. But the animal "did not like to be touched," "seemed fearful, paying close attention to noises," and thus stiffened up, assuming the typical position when sensing danger, keeping "its head very high when disturbed or frightened" with its neck and ears straight. However, it would over time get used to the sight and noises of crowds: the journalists who reported its fears in December would no longer do so in April—either for want of attention or rather because the giraffe overcame them, demonstrating its exceptional individuality. Meanwhile, the novelty had worn off for the inhabitants of Marseilles, who were less eager to see the animal (Salze 1827, 76, 82).[6]

However, the naturalists who would gather around to examine it before its potential death annoyed it. The giraffe was thus immobilized several times by ropes and, later, touched from its rump to its ears so they could measure it, pat it, and draw its portrait. She grew anxious when scientists showed interest in her head, which it shook and "wouldn't let [them] touch easily." It would try to bridle when feeling its lips lifted and its mouth ajar, which it stubbornly closed so tightly that it proved "pretty hard to count the incisors in the lower jaw" and "was impossible to have a notion" of its molars. It would resist even more when these men tried to feel its ears, horns, and a protuberance at the back of its head, as it feared an attack from a predator. "As the animal wouldn't let men touch this part of its body, nothing can be said about it," and they were thus reduced to approximate measures. On other occasions, it was "propelled to flee" to record its movements. It would "hurry, get carried away," thus getting frightened, and would quickly run out of breath. It would otherwise get so "annoyed" when they tried to observe its defence mechanisms that it would rear to scare away the troublesome men, as giraffes do with predators. At other times, it would stretch its neck, raise its head, and fully stick out its tongue to grab at branches which were waved in front of it and then slowly withdrawn to measure the tongue's growth and observe its patterns. It would then feel its tongue grasped and touched—which must have initially frightened and then annoyed it (Salze 1827, 77–9; Geoffroy Saint-Hilaire 1827, 216–220).

The giraffe was also stressed out when the two worlds would not understand one another.[7] It would often have to stoop to catch the branches purposefully put on the ground to expose the strange posture that humans thought to be usual—since they considered eating on the ground as normal—and easy for the animal: "first it moves one of its front legs a little aside, then the other, before doing the same game over again; thus, it only consents to bend its neck and put its lips and tongue on what is offered after several such attempts." In fact, it would be so anxious that it would only go for mimosa, which it loved, as this posture is unsafe in the wild, given predators that may attack the giraffe's bent neck. Giraffes only use it to drink quickly, swallowing 15 litres of water at a time, but never to eat, not hesitating to go without food for three or four days and replacing it with leaves. It would also be physically hindered, taking time and precautions to spread its front legs, contract its rump, stick out its shoulders, and stretch its neck. The repetition of such a position could not but disturb the physiological mechanisms bound to control the strong variations of its blood pressure, as its head repeatedly dropped down six feet under its heart (Geoffroy Saint-Hilaire 1827, 214–15; Mitchell 2006).

With food acting as one of the most important points of the encounter, this giraffe would constantly be invited to sample tastes—and the incomprehension was mutual for a time. It would "more or less" smell, but not touch, European fruits and vegetables (even the fresh ones, though rare and precious for a lot of Western people). "It took some salt but threw it away," although horses and cows delighted in it! It would only swallow "small quantities" of the bread men eat abundantly! It would even turn down exotic fruits, though it was assumed these would appeal to its African origin! It did not care for green meadows—still scarce at the time, and much longed for by European herd animals—since for giraffes, and especially for females, grazing weeds meant adopting a dangerous position! It would barely graze the abundant ash tree, and slightly more so the lime and the cherry trees (rarer and more precious), but "would always come back with pleasure" to the common yew and the thuja, though they were sour and bitter!

Since its capture, it had actually been drinking camel milk, then cow milk—less rich than that of giraffes, but belonging to the ruminant group, and thus easily digestible. This milk became an important element of its psychological stabilisation and physiological survival. "In general, it wouldn't like cold milk" and thus would need it to be tepid—just extracted from the udders—and would hence still be attached to suckling, in the same way many giraffe calves today are bottle-fed and break away from the habit much later. It would thus force men to adapt and to pay attention, but it would nevertheless eat more and more grains and maize (a rare and expensive product), which it "would like better than anything else," and "always eat greedily when served alone." It would even happen to "pick it out, grain by grain, among barley," bran, and broad beans, which it was also given to eat since they were cheaper. And since it was "pleased" by the cows, it started eating the hay it previously refused in February, wanting to imitate them, and partially adjusting its world to others' (Salze 1827, 73–6, 83).[8]

4. From Aix to Paris, the Torment of the Trip

The giraffe's good behaviour triggered the expedition to Paris by hoof from May 20th. It adapted and got accustomed to the rhythm of daily walks. As soon as the 24th, once it "saw the cows preparing to leave, it also decided to leave without waiting anymore for its groom's injunction." It thus reinforced its link with the other animals, achieving "more familiarity" thanks to its contact with men. It became "docile" and "perfectly obedient," to the point that only two mahouts were needed to hold the ropes. It got used to eating and drinking outside and in public, whereas in Marseilles, it would only do so in the peace of the stable (Geoffroy Saint-Hilaire 1827, 214, 219).[9] This is because the giraffe perceived, heard, and felt other animals along the way: it could stare at and smell more-or-less peaceful mules, horses, and oxen on yokes brought to a standstill and placed on the sides of the road by the policemen of the convoy. Hurried crowds—noisier and more turbulent than those in Marseilles during the previous months—would worry it again, causing it to stiffen up and to raise.

The giraffe would mainly stress out along the way in cities with large populations. In Aix, it entered through the narrow streets, seeing many people rushing along its sides or hurrying to buildings' upper floors to see it better, to stretch out the windows and wave. It would hear them shout and applaud around its head, its most sensitive zone. It would always be kept in the storeroom of a tourist hotel but was led outside to fulfil the curiosity of the huge crowds, thus avoiding incidents on departure. It had to stroll twice in one day in Aix—and six times in Lyon—through the streets, the main alleyways, and squares, to allow the greatest possible number of people to see it. In Aix, where "the crowd of curious people was incredible," it could perceive something like "a single clamour so universal, long-lasting and loud it was the brouhaha" as soon as it left in the morning. It felt accompanied by waves of exclamations, explosions of astonishment, and was surrounded by disorganized men: policemen tried to surround and protect the procession from the curious persons who rushed and jostled.

It would not bolt, reassured by the cows that it could see in front of it, compelled by the strong pressure of the tight ropes held by the mahouts, and eventually attracted by the foliage of the avenues and of the squares that were taken on purpose to show how it grazed, offering an upright show that could be watched from afar without pushing. It focused its attention on finding good leaves, forgetting the stress of sounds and movements at its feet. It would also skim the flowers on window sills or the grass grown on lean-tos and even lick a few hands coming its way. It did not seem be "astonished by the crowd that rushed to its feet." This invited its conveyors to expose it, and spectators to become agitated, even though it was indeed worried and stressed out. But it refrained from expressing any of this, like many herbivores do to avoid enticing predators, except for its straight neck and head held up high, a position then mistakenly thought to be the sign of "majestic" calm.

In Lyon, the giraffe panicked during its last parade on Bellecour square. As a precaution to the large popular influx on that morning, the 9th of June, the cows were replaced by a cavalry detachment meant to open the way: the giraffe was thus deprived of its main anchor. "Curious people having rushed to its sides, the animal got scared," as it probably mistook them for predators. It ran off, forced three of four mahouts to drop the ropes, and scared off the horses which ran into the onlookers. It accelerated when it saw the latter trying to avoid its way, perceived the policemen's agitation as they tried to calm down the population, and heard people running behind it to see it galloping, supposing the spectacle to be a planned show. It turned around the statue of Louis XIV before stopping, exhausted by the run and by the weight of the fourth mahout, who still held a rope, but also probably calming down as the crowd slowed (Grognier 6th and 7th June, 9th July 1827).[10]

The giraffe was so stressed out that Geoffroy Saint-Hilaire figured it out: he mentioned the "weariness of representations," like in Aix, where "the population has been insatiable and the giraffe got more tired when resting than because of its days' walks." Thus, the decision was made to cross the Morvan after Beaune in order to avoid other big cities until Paris.[11] The giraffe arrived on June 30th in better shape, thanks to a more gradual physical adaptation, and very calm (it was only led by two mahouts). It was thus led without the cows—deemed too common—to the royal reception in Saint-Cloud, behind two horses ridden by both museum professors, and surrounded by its caretakers, policemen, and a crowd either assembled on its sides or behind it, noisily but harmlessly welcoming it on its way in and out. In front of the yard, it still had to perform for Europeans, who got a (limited and incorrect!) idea of its behaviour. It accelerated under the traction of its caretakers, even though giraffes never trot. It smelled precious petals—a token to the rare animal it was—and as it tried to taste them, had to bend down.

They probably took advantage of the giraffe's apparent calmness, like during its first two-hour walks in the fenced yard of the botanic school at the Jardin des Plantes, where it would hear the shouting crowd gathered at the gates—more than 10,000 people on the day after its arrival. After Saint-Cloud, it went out less often to ensure its rest and avoid indigestion from the flowers brought by the public (who tried to mimic the king and draw the animal closer) and which it greedily ate, as giraffes do (Grognier 15th July 1827; Ferlus 1827b).[12]

The giraffe went out of fashion after the fall of Charles X in 1830. Documents and information on the animal became scarce (Bernard 1842, I, 41–3; Boitard 1842, 442). It got used to the Parisian oceanic climate, to the presence of "popular" neighbours in the rotunda—especially elephants and wild animals with their disturbing smells and roars—and to the death of some of the cows. In 1839, it welcomed another giraffe and could eventually communicate through infrasound with a fellow creature—something it had not been able to do since it left Alexandria, as it was reduced to more-or-less interpret the sounds of other species. Frédéric Cuvier, first ranger and then director of the zoo, often saw it "happily galloping in the alleys of the Jardin des Plantes": this suggests that its exceptional temper was not affected. It grew up to 16 feet—a considerable height for a female specimen—and died at 20 years old in 1845 from phthisis, probably from the milk of one of its cows. Painter Delacroix wrote down in his diary that its death was as unnoticed as its arrival was brilliant (Geoffroy Saint-Hilaire 1842, V, 3; Hartenberger 2010, 121–4).

We shall conclude by stressing the difference between this experimental work and the literary undertakings of animal biographies. Resorting to a scientific method, this approach rejects anthropomorphism and, with a sense of humility, only deals with aspects that were evoked or suggested by the documentary sources. It does not pretend to add or invent elements but states evidence that can be inferred from reliable sources, a common process among historians. By focusing on the animal perspective of a journey, and by casting aside human lenses, we stand far from the human vision of a festive exposure and are confronted with a more delicate encounter between two worlds, with its own fears, misunderstandings, and ways of learning and adapting. These were more important for the newly arrived animal but were just as real for the human beings. They got scared sometimes, excited often, and were at a loss most of the time. They took the time to come, to look, to wonder, and to react, thus fuelling their imagination, their speech, and their texts with this "giraffe world."

Notes

1. This chapter is a reworked version of an article originally published in French as "Une giraffe à la ville (1826–1827). Écrire la rencontre de son point de vue." *Revue d'histoire du XIXe siècle* 54 (2017): 47–62.
2. The letters from Villeneuve-Bargemon and Étienne Geoffroy Saint-Hilaire are available at the Archives Départementales des Bouches-du-Rhône (now ADBR): 4 T 53.
3. ADBR, 4 T 53, Open Account, 10th November 1826, letters from the prefect, 8th, 14th and 18th November 1826, 12th February 1827.
4. ADBR, 4 T 53, letter from the prefect, 12th February 1827.
5. ADBR, 4 T 53, letters from the prefect, 3rd and 19th March 1827 (quote), 28th April 1827, June 1827.
6. ADBR, 4 T 53, letters from the prefect, 12th February and 28th April 1827, from the museum, 4th juillet 1827.
7. The verb "stress out" has been used for about 30 years in ethology, not necessarily with respect to animals' pain but to address a wide range of emotions, such as fear or anxiety, and of reactions such as flight, that are triggered by external pressure.
8. ADBR, 4 T 53, Expense reports, November and December 1826, letters from the prefect, 12th February 1827.
9. ADBR, 4 T 53, letters from Étienne Geoffroy Saint-Hilaire, 21st and 24th May, 2nd June 1827.
10. ADBR, 4 T 53, letters from the prefect, 23rd June 1827, from the deputy prefect of Aix, 23rd June 1827, from Saint-Hilaire, 24th June 1827; Archives Municipales de Lyon, 1 I 65, letter from the Mayor, 6th June 1827.
11. ADBR, 4 T 53, letters from Saint-Hilaire, 23rd and 24th May 1827, 2nd and 5th June 1827.
12. ADBR, 4 T 53, letters from the museum, 4th July 1827, de Saint-Hilaire, 12th July 1827.

References

Allin, Michael (1999). *Zarafa*. London: Headline Review.

Baratay, Éric (2012). *Le Point de vue animal. Une autre version de l'histoire*. Paris: Seuil.

Baratay, Éric (2013). *Bêtes des tranchées. Des vécus oubliés*. Paris: Cnrs éditions.

Baratay, Éric (2017). *Biographies animales. Des vies retrouvées*. Paris: Seuil.

Bashaw, Meredith, et al. (2001). "A Survey Assessment of Variables Related to Stereotypy in Captive Giraffe and Okapi." *Applied Animal Behaviour Science* 73, no. 3: 235–47.

Bernard, Pierre, et al. (1842). *Le Jardin des plantes*. Paris: Curmer.

Bezançon, Rémi (2012). *Zarafa*, film DVD 20th Century Fox.

Boitard, Pierre (1842). *Le Jardin des plantes*. Paris: Dubochet.

Cowie, Helen (2014). *Exhibiting Animals in Nineteenth Century Britain*. New York: Palgrave.

Dagg, Ann (2014). *Giraffe: Biology, Behaviour and Conservation*. Cambridge: Cambridge University Press.

Danto, Arthur (ed.) (2013). Special issue "Does History Need Animals?" *History and Theory* 52, no. 4: 1–167.

Dardaud, Gabriel (2007). *Une girafe pour le roi*. Bordeaux: Élytis.

Ferlus, L.-D. (1827a). *Nouvelle notice sur la giraffe*. Paris: Moreau.

Ferlus, L.-D. (1827b). *Dernière notice sur la girafe contenant la relation de son voyage à Saint-Cloud*. Paris: Moreau.

Geoffroy Saint-Hilaire, Étienne (1827). "Quelques considérations sur la girafe." *Annales des sciences naturelles* 11: 210–23.

Geoffroy Saint-Hilaire, Étienne, et al. (1842). *Histoire naturelle des mammifères*. Paris: Belin.

Ginzburg, Carlo (2012). *Threads and Traces: True False Fictive*. Berkeley: California University Press.

Grognier, Louis-Furcy (1827). *Gazette universelle de Lyon*, 14 february, 27 may, 6, 7, 10, 14 june, 9, 15 july.

Haraway, Donna (1991). *Simians, Cyborgs and Women: The Reinvention of Nature*. New York: Routledge.

Hartenberger, Jean-Louis (2010). *Grandeurs et décadences de la giraffe*. Paris: Belin.

McGrew, William (2010). "New Theaters of Conflict in the Animals Culture Wars." In *The Mind of the Chimpanzee*, edited by Elizabeth Lonsdorf, et al., 173–5. Chicago: Chicago University Press.

Mitchell, Graham, et al. (2006). "The Origin of Mean Arterial and Jugular Venous Blood Pressures in Giraffes." *Journal of Experimental Biology* 209: 2515–24.

Robbins, Louise (2002). *Elephant Slaves & Pampered Parrots*. Baltimore: John Hopkins University Press.

Salze (1827). "Observations faites sur la Girafe envoyée au Roi par le Pacha d'Égypte." *Mémoires du Muséum d'Histoire Naturelle* 14: 68–84.

Whiten, Andrew (2010). "A Coming Age for Cultural Panthropology." In *The Mind of the Chimpanzee*, edited by Elizabeth Lonsdorf, et al., 96–8. Chicago: Chicago University Press.

2 Structures of Captivity and Animal Agency

The London Zoo, ca. 1865 to the Present Times

Violette Pouillard

1. Historiography and the Return of Zoo Animals

Animals have for a long time been excluded from the historiography concerning zoos, a fact which could appear paradoxical for institutions whose essence is based on their exhibition, if it were not for their marginalization from most of historical narratives. Indeed, this exclusion reflects a broader tendency in academic fieldwork, which favours the use of animals as clues to understanding cultural, scientific, social, or political issues (Hribal 2007; Baratay 2012, 17–21). Certainly, for decades, some historians of the environment and rural historians have considered interactions between animals and humans, but usually they focus on the population or species levels, and thereby privilege a long-term perspective (Delort 1984; Isenberg 2000). Moreover, works of cultural history concerning attitudes towards animals, addressing the dialectic between cruelty and compassion at play in their treatment, have evoked their very condition, but not as a topic in and of itself (example: Thomas 1983).

The same lack of interest permeates zoo studies. To be sure, the condition of captive animals was frequently so notable that historians cannot completely avoid mentioning it, but when they do, it is mostly in passing. This general observation must be nuanced in at least two respects: a few zoo studies have paid attention to the condition of animals in order to shed light on the materiality of the institution (Baratay and Hardouin-Fugier 2002), while several historians of culture have addressed the zoo as a microcosmic place which embodies the power dynamics over the 'exotic' alterity in an imperial frame, thereby shedding light on the asymmetries at play in human-animal relationships (Ritvo 1987; Rothfels 2002).

It is only recently that historians of animals have purposefully integrated animals as individuals into historical narratives, thereby considering them for what they are, i.e. sentient, suffering, and acting beings. This "*animal* history" (Baratay 2015) has highlighted anthropozoological dynamics and adaptations, as well as animal resistance, human misunderstandings or indifference, and the influence of all this on historical processes (Baratay 2012; Nance 2013, 2015). One example in the subfield of the history of zoos is Andrew Flack's recent work on the influence of the perception of Misha, a polar bear who arrived at Bristol Zoo in 1979 and displayed "'unnatural' behaviour." Flack has demonstrated how visitors' perceptions, protests by animal rights organizations, and negative press coverage have contributed to the reshaping of acquisitions policies, as well as display and management practices, at Bristol and in British zoological gardens more generally (Flack 2016). However, lots of these works, which aim to rehabilitate animals as a driving force in the history of anthropo-zoological relationships, tend to focus on dynamics and the power of agency, thereby marginalizing the daily animal experiences, largely determined by the power asymmetries which characterize most human-animal relationships (Flack 2018).

In line with the animal turn, this chapter will focus on animal experiences and human-animal relationships in the London Zoo, one of the oldest zoos in Europe, founded in 1828, which was and has remained a renowned institution in the keeping of wild animals, despite its urban and physically limited location. The purpose is mainly to examine the "ambiguities" and "contradictions" which permeate zoo studies, as the zoo appears as not only the site of an implacable separation embodied by the captivity and exhibition of beings which are therefore designated as fundamentally different from ours but also as a place of empathic rapprochement with wild animals (Estebanez and Staszak 2012; Pouillard 2015; Flack 2018, 19–20, 180, 187). The discussion aims to address this issue from below, beyond human discourses. There are many ways of accessing the condition of captive animals, from the quantitative, somewhat dry study of animal longevities, to qualitative approaches such as the biographical (Baratay 2017; Pouillard 2019a, 2019b). Here, particular attention will be devoted to the physical environment of captivity as an indirect way to understand how animals experienced the zoo and if and how they were allowed to influence both their own destiny and the history of these institutions.

The twentieth century seems a particularly propitious time to test these potential reconfigurations, as the "conventional, progressivist, institutional histories of zoological gardens," written by zoo managers themselves or influenced by them—for example in commemorative volumes—have defined the second half of the century as an era of changes, during which "zoological gardens have evolved . . . into something . . . called 'conservation centers'" housing animals who are "no longer the victims of imperial contest; [but] rather, . . . the treasured lucky few" (Rothfels 2002, 24–5). However, as this contribution pays attention to architectural structures of captivity, it is necessary to look back to the nineteenth-century-buildings still in use then. Among the hundreds of different species kept in zoos, the case of primates highlights particularly clearly the tensions and evolutions regarding the confinement of animals. Indeed, being among the most popular captive animals due to their taxonomic proximity with humans, their lives have been marked by privileged interactions with, and attention from, both zoo managers and visitors. They were also valued for their exoticism, a Western construction imposed as a norm, referring to a fascinating alterity, distant yet domesticated, close yet inherently inferior. The display of exoticism, largely modelled on imperial geography (Staszak 2008), went hand in hand with the mass appropriation of tropical animals, rendered fragile by their import into the temperate Western climate. So the fate of primates also illuminates the cracks of human policies marked by inflexibility and hubris.

2. The *New Monkey House* and the "Consumption of the Exotic" (Jones 1997)

In the London Zoo of the late nineteenth and early twentieth centuries, most primates were housed in the *New Monkey House*,[1] a building inaugurated during the Victorian era (1865) and long considered exemplary by foreign zoo managers.[2] Erected "on one of the best sites in the Society's premises" and made of iron and glass in accordance with the latest technical innovations, it offered no outdoor access but aimed to provide animals with "as much light and sun as possible," in opposition to the former "closed window-less building [. . .] long condemned as objectionable in many respects."[3] Access to light had become a critical issue in the nineteenth century. Zoo managers had come to realize that primates, who were locked inside to protect them from cold, languished under such conditions. This fact must be related to confinement and its associated deprivations, including—as research would demonstrate in the twentieth century—a lack of access to UVB rays, inducing vitamin D deficiencies and consequently behaviour and health troubles, such as metabolic bone diseases (rickets and osteomalacia) (Phillipson 1976, 3; Hosey, Melfi, and Pankhurst 2013, 178–9, 454–5). Ironically, these issues could not be solved by keeping animals in well-lit buildings providing light but not UVB rays.

The designers of the *New Monkey House* also intended to offer the monkeys better social conditions. Many primates—as mostly gregarious animals with complex social behaviours—were housed in "large central compartments" (Mitchell 1929, 137), in which they could benefit not only from an increased freedom of movement but also from interactions with their conspecifics (Image 2.1).

However, the significant turnover which characterized captive primate populations at the time certainly impacted their social relationships, as later underlined by veterinary and ethological studies, explaining that the introduction of new individuals into captive groups induces stress and is "often accompanied by initial aggression" (Hosey 2005, 119; see also Demontoy-Bomsel and Berthier 1990, 107). Furthermore, the encyclopaedic paradigm characterizing the zoo impacted every inch of management

Image 2.1 The *New Monkey House*, ca. 1905.
Source: Courtesy John Edwards, see: Edwards 2012, 28.

practices. In zoos as well as in natural history museums, the European scientific and imperial elites subjugated the irreducible abundance and chaos of the natural world by appropriating, classifying, and exhibiting series of specimens, a process which resulted in the importation of animals of hundreds of different species (Pratt 2008, 15–66). Many individuals of the *New Monkey House* were consequently placed in "small separate compartments" against the walls, exhibited like in a living museum (Image 2.1). Moreover, sick animals, and others considered to be in excess, were installed in a "spare compartment," later described by the secretary of the Zoological Society of London as "dark, badly ventilated, and overheated." The initial intention to provide animals with more light and social contacts was thus rapidly overtaken by the fever of appropriation, condemning many monkeys to live a life of solitary confinement in dark surroundings (Mitchell 1929, 137–8 (quotes)). Some improvements were introduced in 1906, when, as a result of the "marked mortality" from tuberculosis during the previous year, the infrastructure was cleaned and renovated and the wooden cages against the walls were removed.[4]

Another crucial attribute of the building resulted from the primacy given to visual access to animals, a central aspect in the theatricality of the zoo, producing an "overexposure" of the captives (Acampora 2005, 71). Also influenced by spatial constraints, primates were thus not separated from the public by a double barrier fencing system but only by wire mesh (Image 2.1). On a physiological level, this physical proximity, which characterized the zoo until the mid-twentieth century, favoured the contagion of zoonoses such as tuberculosis, especially because primates are susceptible to most human diseases (Demontoy-Bomsel 1986, 420). On a behavioural level, it forced the captives, especially those installed against the walls, to experience a very limited degree of control over the interactions with visitors. Several ethological studies highlight the implications of such a proximity in terms of behavioural enrichment, as contacts with visitors, keen on providing animals with food, have an occupational effect for the captives. Yet several works concerning the contemporary zoo—in which animals are provided with much more intimacy—"have also shown results which are best interpreted as indicating that the presence, and particularly the behaviour, of unfamiliar people [. . .] is stressful to [the] animals" (Hosey 2008, 110–16, 121. See also Hosey 2005, 112–15; Hosey, Melfi, and Pankhurst 2013, 239, 264–6, 466, 474–82, 549). Finally, this proximity encouraged direct violence towards primates—another facet of human-animal relationships at the zoo, favoured by the conceptual and physical dominion of animal bodies that zoological gardens embody and allow for (Pouillard 2019b). Zoo managers reacted by attaching signs to the central cages requesting the public to refrain from "irritating the monkeys" (Edwards 2012, 28). Many monkeys, probably infuriated by their captive conditions and the presence of visitors, resisted, in the sense suggested by Susan Nance. Nance uses the notion of resistance not as if animals (in her study, circus elephants) "understood, endorsed, or resisted the world of show business as a human cultural or business practice," i.e. not in a political sense. She argues that animals are capable of resistance insofar as they "rejec[t] the conditions of their experiences" (Nance 2013, 10. See also: Hribal 2010). Here, the captives of the *New Monkey House* fought back: in 1913, additional wire guards were installed around the central cages in order "to prevent children being bitten by the animals." Among 83 "accidents" involving visitors at the zoo in August and September 1922, most of them were "slight bites from monkeys," and of 71 bites inflicted by the zoo animals from January to September 1923, 34 were those of monkeys.[5]

Sanitary, mental, social problems caused by the lack of direct access to fresh air and sunlight, by inappropriate feeding, overcrowding, and human-animal proximity, left a deep impact on the lives and bodies of captive animals, especially as several aforementioned problems went hand in hand with chronic stress and consequently reduced resistance to contagious diseases (Pouillard 2019b, 74–133; Kagan and Veasey 2012; Hosey, Melfi, and Pankhurst 2013, 237). A statistical study based on the London Zoo registers of animals reveals that the average lifespans in captivity of primates who entered the zoo between 1865 (the opening of the building) and 1926 (the year of its destruction), and who died at the zoo, i.e. 6,043 individuals, was to around 433 days, or one year and two months.[6] This average longevity figure is only indicative, especially as it represents longevities *in captivity*, which, except for the very few animals born at the zoo, are not longevities *per se*. It nevertheless provides us with a good indicator, because the vast majority of primates were captured at an early age; either they were transferred subsequently to the zoo, where they arrived young, or they had passed through other institutions or private households, a sojourn which had already killed a proportion of them (Pouillard 2019b, 25–73). We can therefore interpret this figure as an indication of shorter average lifespans due to captivity, especially when compared to figures of longevities provided by syntheses, which give us estimates—although it is difficult to understand what they cover, from life expectancy to upper longevities—which amount to several years; 5 years for Cheirogaleids, a maximum of 12 to 25 years for New World monkeys, 20 to 31 years for Cercopithecids (Macdonald 2010, 286–407). This shortened life of primates led to significant turnover, and, during the building's six decades of existence, around 130 primates on average were imported to the zoo annually.[7] This rhythm reveals the democratization of exotic wildlife permitted by the scale of European expansionism, as well as by the associated globalization of the animal trade and circulation (Rothfels 2002; Robbins 2002; Murray 2007, 115–17; Grigson 2016). This mass appropriation was encouraged by the zoo which participated in the "consumption of the exotic" by not only offering commodified animals for imperial, scientific, and recreational contemplation (Jones 1997, 7) but also literally consuming hundreds of animal bodies a year.

The industrial commodification of animals seems to be contradicted by certain individual trajectories. Several primates, imported at a young age and entrusted to the care of the staff, aroused deep interest and affection. A famous example in the late nineteenth century concerns the chimpanzee Sally, aged between two and four when she arrived at the zoo in 1883. Trained to count and to eat with cutlery like lots of her conspecifics (Baratay 2017, 127–46), she escaped confinement in the *New Monkey House* and became a local celebrity.[8] Yet such cases, usually concerning anthropoid apes, were rare and did not prevent these animals from experiencing the general constraints of captive life, such as confinement, lack of control over interactions with humans, or deprivation of social contact with conspecifics. The longevity of Sally, who died after eight years at the zoo, was considered by contemporaries to be a "remarkable exception."[9] It is also difficult to assess if the training sessions imposed on many chimpanzees were beneficial or detrimental to their well-being (Pouillard 2019b, 123–4).

3. The New *New Monkey House*, or the Hygienist Turn

The era of the Victorian *New Monkey House* came to an end in 1926. As a result of the import policies and structural sanitary problems, a severe tuberculosis epidemic broke out in the house, forcing the London Zoo managers to hasten the long-postponed construction of a new building.[10] At the time, due to the deterioration of the 1865 edifice, the housing of primates was quite chaotic: they were "housed under bad conditions in many parts of the Gardens," most of them being deprived of "free exposure to open air and sunlight" and forced to experience "no heat or . . . a monotonous, relatively low temperature from a hot-water system."[11]

The new monkey building opened in 1927 was constructed in line with a provisional *Experimental Monkey House*, erected around two years earlier, which testified to the increasing rationalization associated with the management of primates, and of animals more generally, enforced by the secretary of the Zoological Society of London (1903–1935) Peter Chalmers Mitchell. Mitchell, trained as an anatomist, developed networks both in the medical world—he was close to Leonard Hill, his colleague at the London Hospital Medical College, and a fervent advocate of hygienism—and in the zoo world—he developed a friendship with Carl Hagenbeck, the famous animal-dealer who in 1907 opened a zoo in Stellingen, near Hamburg, characterized by fence-free, open-air enclosures. These relationships contributed to his aroused enthusiasm for the "cold air school," which intended to give animals access to fresh air, instead of confining them permanently in over-heated dens (Mitchell 1929, 189). Furthermore, the rise of international conservationist concerns towards African fauna from the late nineteenth century onwards could have played a part in Mitchell's attempts to rationalize the management of primates, as he presided from 1923 to 1927 over the Society for the Preservation of the Fauna of the Empire, an influential lobbying conservationist organization. Aided by the healthy financial situation of the Zoological Society of London, he was able to implement an ambitious architectural and management programme (Edwards 2004; MacKenzie 1997; Pouillard 2019b 167–235).

Among these measures, the development of the *Experimental Monkey House* aimed to ensure "an installation as perfect as possible in every detail for the housing of all the Primates." The main improvements tested in the structure included the provision of fresh air for the captives via access to inner and outer cages, of modulated heating via a supply of warmed shelves and plates, and of additional, artificial exposure to UV rays through the intercession of quartz incandescent lamps,[12] thereby solving the hygienist's triple concern for air, temperature, and sunlight.

The abruptness of the 1926 tuberculosis epidemic, and a subsequent financial assessment, convinced the zoo administrators both to hasten the construction of a *New Monkey House*—drawing from the *Experimental Monkey House* as a prototype—and to construct only two-thirds of the planned building.[13] Due to the failure to execute fully the guiding principles of the edifice as well as to its structural characteristics, the desired rationalization of the new *New Monkey House* was quickly jeopardized. Cages were separated both from each other and from visitors by wire mesh. This design deprived the captives of any physical, visual, acoustic, and olfactory insulation, which provoked tensions and conflicts and increased stress levels, as later explained by the curator of mammals:

> Males became jealous when neighbouring pairs copulated and their displays often resulted in copulations being broken off to enable counter-displays to be made. Often this caused a chain reaction until the whole house was filled with a deafening hubbub and all the animals were in a state of high tension. Several mothers were unable to cope and failed to nurse their new-born normally. Fighting between adults and juveniles was frequent.
>
> (Toovey and Brambell 1976, 212)

Beyond these dantesque social conditions, this option still allowed for a certain degree of proximity with the audience, from whom monkeys were separated by a barrier located around 90 centimetres from the wire mesh—around December 1931, a six-year-old child "had a finger badly bitten by a monkey"—while some visitors were authorized to see the animals behind the scenes. These contacts, and the associated risks of contagion, were neutralized in 1936, in the anthropoid section only, by the erection of a glass screen and the prohibition of visits behind the cages.[14] Finally, the hygienist devices themselves did not stand the test of time. The electrically heated plates on which the primates could sit began "to short circuit and overheat as they were penetrated by urine" (Toovey and Brambell 1976, 212). The ideal of providing primates with free outdoor access rapidly evaporated due to the desire of satisfying the visitors' omniscient gaze, highlighting the prevalence given to recreational purposes over consideration for the needs of the animals:

> From the public's point of view, it was frustrating to go down inside the house if the monkeys had chosen that moment to go outside. It was even more frustrating to make the trek round to the corresponding outside cage, only to find that the animals had gone in again. Of course the public complained, and it became the practice to decide for the monkeys whether it was a good day for all to be in, or for all to be out. Since conscientious keepers always made this decision on the safe side, there was a tendency for the animals to be kept indoors on marginal days.
>
> (Toovey and Brambell 1976, 212)

If the new building had been erected to avoid indoor confinement, management practices continued to perpetuate these conditions. Due to these combined failures, the rationalization was embodied mainly by the bareness of the cages, made of concrete and tiles, "cramped" and deprived of enrichment except for a few shelves and perches (Brambell and Mathews 1976, 152) (Image 2.2).

Such a hygienist turn, here examined on an architectural level, generated a slight increase in average primate longevities in captivity: during the more than four decades of the building's use (1927–1970), they amounted, for animals who died at the zoo (n = 2,125), to around two years and five months.[15] The associated impoverishment radically amputated the animals' *Umwelt*—this Uexküllian notion meaning the subjective world of animals, constructed by their perceptions and actions and thereby populated with meaningful objects (Uexküll 2010 [1934])—and in doing so, exacerbated the stress and aggressiveness, as was evoked by the managers in the aforementioned excerpt, which certainly in turn impacted mortality rates.

Image 2.2 Indoor cage, 1927 Monkey House.

Source: © Zoological Society of London.

4. The *Michael Sobell Pavilions*, or the Conservationist Turn

In 1972, the London Zoo inaugurated a new primate building, the *Michael Sobell Pavilions for Apes and Monkeys*, which had been designed for the display of around 13 species, in contrast with the more than 30 exhibited in the previous house, a number which was subsequently further decreased.[16] This mitigation of the encyclopaedic paradigm must be related to the new function attributed to primates: they must henceforth breed that the trade in animals, which was undergoing expanding criticism and restrictions in the postcolonial era, could be bypassed (Baratay and Hardouin-Fugier 2002, 223–30). Zoos, rather than removing primates from their environment, should hence produce them through captive breeding, which implied that the captives should be housed in social groups.[17]

This transformation must also be associated with increasing attention paid to the occupational and psychological needs of animals as a reaction to boredom and the behavioural troubles induced by the bareness of the captive environment, exacerbated by the hygienist turn, as well as to the unprecedented development of animal rights concerns (Singer 1975; Kean 1998, 197–214). In the *Pavilions*, hygienist requirements such as easy-to-clean materials, glass separations, and adjustable heating and ventilation systems were intermingled with measures improving animal welfare. For the first time at the London Zoo, primates were provided with effectively free outdoor access during the day (Image 2.3). They were

Image 2.3 The Sobell Pavilions.

Source: © Zoological Society of London.

given an unprecedented range of enrichment devices (such as platforms, climbing frames, aerial structures, etc.) and benefited from activities such as the search for sunflower seeds scattered on the grass ground. Stress induced by captivity was mitigated by the provision of reinforced physical separations with visitors, in order to allow more control over interactions, and of several access routes to indoor and outdoor enclosures, in order to allow better interactions with the conspecifics. Three years after the opening, the zoo managers ensured that "most groups ha[d] become established without difficulty" and that "there [was] no longer the state of tension among the animals which was such a drawback in the old house" (Toovey and Brambell 1976, 217).

However, some managers' comments made generosity sound like stinginess: the "choice of routes" aimed to "discourage the development of repetitive patterns of movement" (revealing the prevalence of stereotypical behaviours) and to "prevent animals [from] being cornered by aggressive companions" (indicating potential conflicts between captive animals unable to escape), while "rubber ropes ha[d] been suspended from the frame to further improve the variety of movement" (attesting to the restriction of mobility) (Toovey and Brambell 1976, 214–15). Primates continued to live in an impoverished *Umwelt* as indicated among other things by the "concern" expressed in 1997 by the Zoological Society of London's Animal Welfare Committee regarding the indoor enclosures, which, "out of public view," were then described as "bare," "which may discourage the animals [from] enter[ing] and therefore deny them privacy."[18]

Microhistorical insights, a perspective which requires further research, shed more light on the experiences of captive primates. The Lowland gorilla Guy, one of the most popular animals in the history of the zoo, wild-born and brought to London as a one-year-old in 1947, lived in forced-isolation for more than two decades, contrary to the gregarious organization of his species. In 1969, "[i]n preparation for stocking the new Primate House," the zoo managers imported a female, Lomie, who was kept in a cage adjoining that of Guy in the 1927 *New Monkey House*, before the two were finally housed together in 1972, when the new *Pavilions* opened.[19] The motivations behind the decision to interrupt Guy's decades-long loneliness seem to be related to breeding goals. Lomie had been imported "as a possible future mate".[20] As was the case for many gorillas in the 1960s and 1970s,[21] Lomie and Guy were forced to live as a couple, another social structure contrary to the gorillas' own. Finally, when it turned out that, after three years in the company of Lomie, Guy had not produced any offspring—a fact which must be related to his sterility (Dixson, Moore, and Holt 1980)—Lomie was sent to Bristol Zoo in order to breed, leaving Guy alone, again.[22] This totemic example demonstrates that the development of captive breeding as part of the conservationist turn did not eradicate the commodification of the captives, considered both as individuals and as representatives of their species and "reservoirs" of genes, i.e. as breeding tools.

5. Epilogue: The *Gorilla Kingdom*

Attention to captive animals as individuals would gradually be given more weight in management policies, and the development of new infrastructures once again embodied and catalysed this evolution. The *Sobell Pavilions* were partially replaced in 2007 by a *Gorilla Kingdom*, which, focused on a single species, represents the nadir of the encyclopaedic trend. It also exemplifies the apex of attention towards the occupational and social aspects of animals' lives in captivity, be it through free access to a planted outdoor island during the day, the unprecedented development of enrichment devices, or through the social environment, as its gorillas live in groups constituted to mimic their gregarious structure.[23]

However, such improvements, like the others conducted before them, cannot avoid relying on a defragmentation of the animals irreducible *Umwelt* in a series of elements (sunlight, temperature, fresh air, social relationships, nutritional requirements, occupational needs, etc.) as a prerequisite to its artificial reconstruction, which consequently cannot be anything but partial. For example, free-living Western gorillas eat a high number of plants and invertebrates (148 species as a minimal average estimation) with seasonal variations, travel long distances each day in search of food, and select certain parts of plants (fruit, leaves, stems, pith, roots, shoots, bark, etc.) (Remis 1997; Rogers 2004). This very evocation of their life cannot itself avoid leading back to impoverishment, as the abundance of natural surroundings and, even more, of the animal uses of them, cannot be easily caught and described by scientific studies (see, by contrast, Bailly 2013). However, it gives us a glimpse into the deep intermingling of hedonistic, occupational, territorial, social, and nutritional aspects in the act of feeding and thereby shows us how difficult it is to offer animals complete experiences when making decisions on their behalf.

Furthermore, the intrinsically omnipotent character of the management of zoo animals cannot escape the imposition of restrictions on the captives, expressed for example by the fact that gorillas cannot choose their conspecifics or emigrate voluntarily from their group, as regularly occurs in the wild (Harcourt 1987, 250, 253; Robbins 2004). Further constraints are produced by the "overexposure" described by Ralph Acampora (Acampora 2005), resulting in the now indirect, but still forced, contact with crowds, the effects of which have been highlighted in terms of stress by several ethological studies (Hosey 2005; Wells 2005). The London Zoo managers recognize this latter impact as indicated by the notice "Do not bang on the glass, as it frightens the animals" attached to the glass of the indoor part of the *Gorilla Kingdom*,[24] further demonstrating that physical violence towards captive animals as a product of their exhibition has now, in the era of increasing physical separation between the visitors and zoo animals, been reshaped under new forms, especially harassment.

Simultaneously, despite the constant reinforcement of devices of confinement for security purposes, zoo animals have never ceased to resist captivity. Several examples indicate that gorillas in their *Kingdom* have not always complied with the docility expected from them. On 13 October 2016, the male Kumbuka escaped from the indoor enclosure through two inadvertently unlocked doors and remained in the keeper-area for more than one hour, where he drank five litres of undiluted blackcurrant juice. Even if he was subsequently "tranquilized" (Press Association 2016), the incident was not without impact. On the short term, it provoked the lockdown of the zoo and reactions of panic among visitors. On the medium term, it gave rise to press reactions concerning the lifestory of the gorilla, born in Belfast on 15 November 1997, rejected by his mother—as regularly happens in captivity—and successively transferred to the zoos of La Boissière (2002), Paignton (2006), and London (2013)—summarizing the animal side of the conservationist turn, which has transformed the life of certain animals (particularly those interesting for breeding purposes) into a kind of itinerant career. Some articles underlined his "agitat[ion]," especially as he was regularly charging at the glass, a display-behaviour which can indeed be identified with aggressiveness, and related it to his attempt to escape. Finally, the incident prompted animal welfare organizations to call for inquiries about the safety and welfare of apes in zoos and made famous conservationists Ian Redmond and David Attenborough come out for the privacy of captive gorillas (Wilms et al. 2011; Meder 1989, 367; Bagot and Myers 2016 (quote); Agence France-Presse 2016). Since then, the London Zoo managers, in an attempt "to make the presence of human visitors less obstrusive to the animals," have placed stickers representing foliage on part of the glass of the indoor enclosure.[25]

If Kumbuka's episode of resistance demonstrates how animal actions and responses have a great potential to reshape management practices, it cannot mask the limits of animal agency as a driving force in the history of human-animal relations marked by a structural asymmetry. Firstly, while the *Gorilla Kingdom* was attracting visitors' attention, many primates were still living in renovated but old encyclopaedic buildings, like the nocturnal primates kept inside the refurbished *Clore Small Mammal House* (1967).[26] At the same time, most reptiles, amphibians, and invertebrates were alike confined without any outdoor access. Secondly, the iconic gorillas themselves cannot escape living in an impoverished *Umwelt* and experiencing stress due, among other things, to the presence of crowds, only mitigated through the foliage stickers.

Therefore, structural aspects of animal captivity, such as the architecture of the zoo, must not be marginalized by the recent focus on animal agency, encouraged by the animal turn. Although the study of zoo buildings appears particularly dry in comparison with research on historical anecdotes which affiliate humans and animals at the zoo, it demonstrates that these material constructions reflecting conceptual frameworks have impacted the lives of animals more deeply than contingent interactions with humans (the latter themselves restricted by these structural constraints). Of course, this approach is too incomplete to access individual experiences and must therefore be complemented by biographical insights (Pouillard 2019a). This multiplication of perspectives will allow us to write a history of animals which offers a representative balance of, on the one hand, the expression of animal agency, and, on the other hand, the restrictions imposed on the latter in a coercive context. Recognizing the breakdown of the human-animal relationship in such environments does not deny anything to the animals. As Susan Nance has demonstrated regarding elephants in the nineteenth-century circus, they "had agency but no real power . . . they had no effective way to challenge circus life broadly," and circus men did not act to give them more power, certainly because this would have endangered their institution and their career (Nance 2013, 97). The same can be said of zoo animals. This helps to explain why their lives appear as a succession of struggles, minor victories, and impediments to the full exercise and actualization of their agency.

Notes

1. This chapter focuses on the general, larger buildings housing primates, yet several of them were installed in infrastructures maintaining other species as well, as, from the early twentieth century on, in buildings dedicated to specific taxonomic groups: the *Ape House* (1902), the *Monkey Hill* (for baboons, 1925), the *Lemur House* (1928, former *Experimental Monkey House*), and the *Gorilla House* (1933).
2. National Archives, Pierrefitte-sur-Seine, AJ/15/845, A. Milne-Edwards, *Rapport sur l'état actuel de la Ménagerie . . .*, [1891].
3. *Report of the Council of the Zoological Society of London . . . 1864*, 9, 15–7 (quotes), *1865*, 14.
4. ZSL Archives, GB0814 PAA. *London Zoo Committee (LZC)*, 10 January 1906; *Report of the Council of the Zoological Society of London, for the year 1905*, 32 (quote), *1906*, 28, *1907*, 40.
5. Monkeys were then housed in several places, but most of them were confined to the *Monkey House*. ZSL Archives, GB0814 PAA. *LZC*, 18 October 1922 (quotes 2–3), 16 October 1923; *Report of the Council of the Zoological Society of London, for the year 1913*, 30 (quote 1).
6. After: ZSL Archives, GB0814 QAA. *Daily Occurrences* 1865–2000.
7. After: ZSL Archives, GB0814 QAA. *Daily Occurrences* 1865–2000.
8. ZSL, GB0814 QAA. *Daily Occurrences*, 24 October 1883; "Sally, the bald chimpanzee." 1891. *The Animal World* 266: 164–5.
9. *Idem* (quote); ZSL, GB0814 QAA, *Daily Occurrences*, 28 August 1891.
10. ZSL, *Report of the Council of the ZSL, for the year 1926*, 32–3, Mitchell (1929, 205).
11. *Report of the Council of the Zoological Society of London, for the Year 1924*, 35.

12. *Report of the Council of the Zoological Society of London, for the Year 1925*, 29–30.
13. *Report of the Council of the Zoological Society of London, for the Year 1926*, 32–3.
14. ZSL Archives, GB0814 PAA. *LZC*, 15 December 1931 (quote); *Report of the Council of the Zoological Society of London, for the year 1936*, 11.
15. After: ZSL Archives, GB0814 QAA. *Daily Occurrences*, 1927–2000. Once again, this figure is indicative, especially as it encompasses longevities of animals housed in the *Monkey Hill*, opened in 1925, and in the *Gorilla House* (1933).
16. *Annual Report 1970*, ZSL, 15, *1992–1993*, 11, *Supplement to the Annual Report 1998*, 18.
17. *Annual Report 1975*, ZSL, 10; Toovey and Brambell (1976, 212).
18. ZSL Archives, *Animal Welfare Committee, Minutes of the Meeting Held on Tuesday 13 May 1997 at Whipsnade Wild Animal Park*, 3.
19. *Annual Report 1969*, ZSL, 19 (quote), *1971*, 11, Wilms (2011). Guy may have had brief contact with the young gorillas Rundi, Tanga, and Reuben in 1960–3 (see n. 9929–31).
20. *Annual Report 1969*, ZSL, 19.
21. Example: Stadsarchief Antwerpen, 1#1872, Van den Bergh to Jones, 29 January 1975.
22. *Annual Report 1975*, ZSL, 9–10.
23. *ZSL Living Conservation 2006–2007*, 10.
24. Personal visits to London Zoo, 17 June 2012, 3 May 2018.
25. Visit to the London Zoo, 3 May 2018, Hosey, Melfi, and Pankhurst (2013, 477) (quote).
26. *Annual Report 2006–2007. Living Conservation*, ZSL, 10.

References

Acampora, Ralph (2005). "Zoos and Eyes: Contesting Captivity and Seeking Successor Practices." *Society & Animals* 13, no. 1: 69–88.

Agence France-Presse (2016). "David Attenborough: Zoos Should Use Peepholes to Respect Gorillas' Privacy." *The Guardian*, October 18. Accessed April 13, 2018. www.theguardian.com/world/2016/oct/18/david-attenborough-zoos-respect-gorillas-privacy-peepholes.

Bagot, Martin, and Myers, Russell (2016). "Escaped London Zoo Gorilla Kumbuka Had 'Difficult Upbringing' and 'Needs Privacy'." *Mirror*, October 15. Accessed April 13, 2018. www.mirror.co.uk/news/uk-news/escaped-london-zoo-gorilla-kumbuka-9048626.

Bailly, Jean-Christophe (2013). *Le Parti pris des animaux*. Paris: Seuil.

Baratay, Éric (2012). *Le point de vue animal. Une autre version de l'histoire*. Paris: Seuil.

Baratay, Éric (2015). "Building an *Animal* History." In *French Thinking about Animals*, edited by Louisa Mackenzie and Stephanie Posthumus, 3–14. East Lansing: Michigan State University Press.

Baratay, Éric (2017). *Biographies animales. Des vies retrouvées*. Paris: Seuil.

Baratay, Éric, and Hardouin-Fugier, Elisabeth (2002). *Zoo: A History of Zoological Gardens in the West*. London: Reaktion Books.

Brambell, Michael R., and Mathews, Sue J. (1976). "Primates and Carnivores at Regent's Park." In *The Zoological Society of London, 1826–1976 and Beyond*, edited by Solly Zuckerman, 147–65. London: Academic Press/The Zoological Society of London.

Delort, Robert (1984). *Les Animaux ont une histoire*. Paris: Seuil.

Demontoy-Bomsel, Marie-Claude (1986). "Contribution à la pathologie des primates." *Recueil de médecine vétérinaire* 162, no. 3: 411–25.

Demontoy-Bomsel, Marie-Claude, and Berthier, Jean-Luc (1990). "Quelques exemples de pathologies liées à l'environnement." In *Erkrankungen der Zootiere. Verhandlungsbericht des 32. Internationalen Symposiums über die Erkrankungen der Zoo-und Wildtiere*, 107–13. Berlin: Akademie-Verlag.

Dixson, A.F., Moore, H.D.M., and Holt, W.H. (1980). "Testicular Atrophy in Captive Gorillas (*Gorilla g. Gorilla*)." *Journal of Zoology* 191, no. 3: 315–22.

Edwards, J.C. (2004). "Mitchell, Sir Peter Chalmers (1864–1945)." In *Oxford Dictionary of National Biography*, edited by H. C. G. Matthew and Brian Harrison, 38, 425–6. Oxford: Oxford University Press.

Edwards, John (2012). *London Zoo From Old Photographs 1852–1914*. London: John Edwards.

Estebanez, Jean, and Staszak, Jean-François (2012). "Animaux humains et non-humains au zoo. L'expérience de la frontière animale." In *Aux frontières de l'animal. Mises en scène et réflexivité*, edited by Annick Dubied, Juliet J. Fall, and David Gerber, 149–74. Geneva: Librairie Droz.

Flack, Andrew (2016). "'In Sight, Insane': Animal Agency, Captivity and the Frozen Wilderness in the Late-Twentieth Century." *Environment and History* 22, no. 4: 629–52.

Flack, Andrew (2018). *The Wild Within: Histories of a Landmark British Zoo*. Charlottesville/London: University of Virginia Press.

Grigson, Caroline (2016). *Menagerie: The History of Exotic Animals in England, 1100–1837*. Oxford: Oxford University Press.

Harcourt, A.H. (1987). "Behaviour of Wild Gorillas *Gorilla Gorilla* and Their Management in Captivity." *International Zoo Yearbook* 26: 248–55.

Hosey, Geoffrey R. (2005). "How Does the Zoo Environment Affect the Behaviour of Captive Primates?" *Applied Animal Behaviour Science* 90: 107–29.

Hosey, Geoffrey R. (2008). "A Preliminary Model of Human-Animal Relationships in the Zoo." *Applied Animal Behaviour Science* 109: 105–27.

Hosey, Geoffrey R., Melfi, Vicky, and Pankhurst, Sheila (2013). *Zoo Animals: Behaviour, Management, and Welfare*. Oxford: Oxford University Press.

Hribal, Jason (2007). "Animals, Agency, and Class: Writing the History of Animals From Below." *Human Ecology Review* 14, no. 1: 101–12.

Hribal, Jason (2010). *Fear of the Animal Planet: The Hidden History of Animal Resistance*. Petrolia: AK Press.

Isenberg, Andrew C. (2000). *The Destruction of the Bison: An Environmental History, 1750–1920*. Cambridge: Cambridge University Press.

Jones, Robert W. (1997). "'The Sight of Creatures Strange to Our Clime': London Zoo and the Consumption of the Exotic." *Journal of Victorian Culture* 2, no. 1: 1–26.

Kagan, Ron, and Veasey, Jake (2012). "Challenges of Zoo Animal Welfare." In *Wild Mammals in Captivity: Principles & Techniques for Zoo Management*, edited by Devra G. Kleiman, Katerina V. Thompson, and Charlotte Kirk Baer, 11–21. Chicago/London: The University of Chicago Press.

Kean, Hilda (1998). *Animal Rights: Political and Social Change in Britain Since 1800*. London: Reaktion Books.

Macdonald, David W. (ed.) (2010). *The Encyclopedia of Mammals*. Oxford: Oxford University Press.

MacKenzie, John (1997). *The Empire of Nature: Hunting, Conservation and British Imperialism*. Manchester/New York: Manchester University Press.

Meder, Angela (1989). "Effects of Hand-Rearing on the Behavioral Development of Infant and Juvenile Gorillas (*Gorilla g.gorilla*)." *Developmental Psychobiology* 22, no. 4: 357–76.

Mitchell, Peter Chalmers (1929). *Centenary History of the Zoological Society of London*. London: Harrison & Sons.

Murray, Narisara (2007). "From Birds of Paradise to *Drosophila*: The Changing Role of Scientific Specimens to 1920." In *A Cultural History of Animals, 5, in the Age of Empire*, edited by Kathleen Kete, 113–34. Oxford/New York: Berg.

Nance, Susan (2013). *Entertaining Elephants: Animal Agency and the Business of the American Circus*. Baltimore: The Johns Hopkins University Press.

Nance, Susan (2015). *Animal Modernity: Jumbo the Elephant and the Human Dilemma*. New York: Palgrave Macmillan.

Phillipson, A.T. (1976). "Principles of Zoo Animal Feeding: Introduction." *International Zoo Yearbook* 16: 1–3.

Pouillard, Violette (2015). "Entre affections et violences. Visiteurs et animaux de zoo du XIXe siècle à nos jours." *Revue semestrielle de droit animalier* 2: 309–25.

Pouillard, Violette (2019a). "Gust (*ca*. 1952–1988), or a History from Below of the Changing Zoo." In *Zoo Studies: A New Humanities*, edited by Tracy MacDonald and Daniel Vandersommers. Kingston: McGill-Queens University Press.

Pouillard, Violette (2019b). *Histoire des zoos par les animaux. Contrôle, impérialisme, conservation*. Seyssel: Champ Vallon.

Pratt, Mary Louise (2008, 2nd ed.). *Imperial Eyes. Travel Writing and Transculturation*. London, New York: Routledge.

Press Association (2016). "London Zoo Gorilla Drank Five Litres of Blackcurrant Juice after Escaping Enclosure." *The Guardian*, October 20. Accessed April 13, 2018. www.theguardian.com/world/2016/oct/20/gorilla-drank-litres-blackcurrant-juice-escaping-london-zoo-enclosure.

Remis, Melissa J. (1997). "Western Lowland Gorillas (*Gorilla Gorilla Gorilla*) as Seasonal Frugivores: Use of Variable Resources." *American Journal of Primatology* 43: 87–109.

Ritvo, Harriet (1987). *The Animal Estate: The English and Other Creatures in the Victorian Age*. Cambridge: Harvard University Press.

Robbins, Louise E. (2002). *Elephant Slaves and Pampered Parrots: Animals in Eighteenth-Century Paris*. Baltimore/London: The Johns Hopkins University Press.

Robbins, Martha M., et al. (2004). "Social Structure and Life-History Patterns in Western Gorillas (*Gorilla Gorilla Gorilla*)." *American Journal of Primatology* 64: 145–59.

Rogers, Elizabeth M. (2004). "Western Gorilla Diet: A Synthesis from Six Sites." *American Journal of Primatology* 64: 173–92.

Rothfels, Nigel (2002). *Savages and Beasts: The Birth of the Modern Zoo*. Baltimore/London: The Johns Hopkins University Press.

Singer, Peter (1975). *Animal Liberation: A New Ethics for Our Treatment of Animals*. New York: New York Revue.

Staszak, Jean-François (2008). "Qu'est-ce que l'exotisme ?" *Le Globe. Revue genevoise de géographie* 148: 7–30.

Thomas, Keith (1983). *Man and the Natural World, Changing Attitudes in England 1500–1800*. London: Allen Lane.

Toovey, John, and Brambell, Michael (1976). "The Michael Sobell Pavilions for Apes and Monkeys." *International Zoo Yearbook* 16: 212–17.

Uexküll, Jakob Von (2010 [1934]). *Milieu animal et milieu humain*. Paris: Payot & Rivages.

Wells, Deborah L. (2005). "A Note on the Influence of Visitors on the Behaviour and Welfare of Zoo-Housed Gorillas." *Applied Animal Behaviour Science* 93, no. 1–2: 13–17.

Wilms, Thomas, et al. (ed.) (2011). *International Studbook for the Western Lowland Gorilla* Gorilla g. Gorilla *Savage & Wyman, 1847*. Frankfurt: Frankfurt Zoo.

3 Atlantic History From the Saddle

The Role of Horses in the Slave-Trading Atlantic World

Rachael L. Pasierowska

> He requested they take his body down the river to this his favourite haunt, and on the pinnacle of this towering bluff, to bury him on the back of his favourite war-horse, which was to be buried alive, under him . . . he was placed astride of his horse's back, with his bow in his hand, and his shield and quiver slung.
>
> (Catlin 1866, 1)

In his letter, George Catlin, the American painter, writer, and traveller, described the retelling of the interment of "Black Bird," the renowned O-ma-haw's chief who perished of smallpox in the nineteenth century. The O-ma-haw people's behaviour in the epigraph of the chapter demonstrates a strong cultural association with the horse as a creature of warfare but also as a being who would allow the recently deceased Black Bird to make the crossing from the tangible world of the living to the "beautiful hunting-grounds of the shades of his fathers"—the intangible spiritual world. For the O-ma-haws, the horse was a creature that denoted both physical *and* spiritual power. As Native American Indians, however, Black Bird and his community did not have a long history with the horse. Indeed, this equine beast did not become a feature of North American life until at least a century after the introduction of the first horses into Central and South America following the Spanish Conquest. However, the horse would soon become an iconic feature of many North American indigenous societies, as depicted in the burial of Black Bird.

This chapter explores the shifting fate of the horse as a subject of inquiry in the Atlantic World, primarily during the era of the trans-Atlantic slave trade. During recent decades, scholars of the Atlantic World have been keen to foreground the experiences of groups of peoples commonly overlooked or not focused on in much detail. This has resulted in historians devoting greater attention to all peoples of the Atlantic world, for instance Africans and Native Americans, and not just white Europeans. It is important to stress, as does the renowned historian of English overseas expansion Bernard Bailyn, that Atlantic history is not merely "an expansion of the venerable tradition of 'imperial' history" (2005, 5). In its most simplistic form, one function of Atlantic history is to both de-Europeanize and de-nationalize the scholarship surrounding the creation of the Atlantic world and its ensuing inter-continental economies. In the words of Trevor Burnard "[p]erhaps the single most important advance attributable to the Atlantic perspective has been its encouragement of the incorporation of Africans and Native Americans into the making of colonial British America" (2009, 121). Yet, while scholars have turned with greater focus to the lives of indigenous peoples and the movement of peoples and slave-produced crops across the Atlantic World, there remain other migrants that continue to receive little or no direct attention: animals.[1] This chapter adopts a cultural history approach to look at the horse as a lens through which one may analyse the changing face of the Atlantic World during the era of African slavery in the Americas. Moreover, this is a circum-Atlantic approach, utilizing the term as defined by David Armitage (2009, 15), or the study of Atlantic World history through the lens of transnational history. Adopting such an approach enables one to better understand peoples and nations that may appear at a first glance rather dissimilar in their view and practices regarding the horse.

Placing the horse at the centre of our focus of analysis foregrounds the multiple ways in which the numerous meanings and symbolic representations of the horse shed light on the system of Atlantic slavery, the migration of Europeans to the Americas, and contact with Native American peoples. The movement of peoples across the Atlantic World enabled this circulation of values and beliefs regarding horses, which were assimilated and transformed by different racial and social groups, and facilitated through pre-existing similarities in beliefs and perceptions of the horse, notably that of a creature that symbolized power and prestige. Looking over large temporal and geographical scales, this chapter purports to show how similar beliefs, interests, practices, and values regarding the horse circulated among historical actors who, at a first glance, appear far removed from one another, despite their connections across the Atlantic World. Three patterns will be identified and studied in detail: (1) the horse as an agent of power; (2) the horse as a disputed symbol of power and prestige; and, finally, (2) the metaphorical representation of the horse in a spiritual light.

This chapter utilizes a wide range of sources—both primary and secondary—from across Africa, the Americas, and Europe. Working with such large geographical and temporal scales requires the historian to employ different research approaches. For the nature of this study, there is no one-size-fits-all; indeed, sources often need to be read against the grain or juxtaposed alongside other materials. Secondary sources come from multiple disciplines, principally history, but also the fields of anthropology, art history, literature, and religious studies. To craft a more nuanced understanding of horses in the lives of people across the Atlantic World,

I bring together different studies in a broader Atlantic World approach, allowing the identification of similarities and differences that are often hitherto overlooked. Alongside the secondary sources, a wide range of primary sources are also employed: while European travel accounts comprise the greater proportion of these, I also look to non-written sources, such as folklore, material and visual culture, and oral history to better explore the voice of the often underrepresented or the exploited indigenous Africans and Native Americans. It is this juxtaposition of sources and histories which allows a more nuanced understanding of horses in the lives of those across the Atlantic World. This chapter, moreover, seeks to move beyond the national boundaries of the Anglo-Atlantic World, incorporating data from both the northern- and southern-hemisphere slave trades to draw analogies and juxtapositions between and across cultures.

1. Horses as Agents of Power

In the year 1493, Columbus made landfall in the Americas marking his second voyage, along with the first equines since their extinction during the Pleistocene epoch. Unlike the pig, which multiplied in abundance in the Caribbean, horses did not adapt with as much alacrity to their new world environs. Moreover, the death rates for horses crossing the Atlantic were particularly high (Crosby 2003, 80). Nonetheless, as a result of continued exports from Spain and breeding in the Americas, horse numbers increased. The investment paid off well, given the central role that horses played for the Spanish in their conquest of the Americas. In an ironical twist, however, equines would later prove greatly important to the very peoples the Spanish sought to dominate.

Before Native Americans incorporated horses into their societies, exposure to these creatures first had to occur. Across the Americas, the horse evoked a mixture of curiosity, fear, and otherworldly sentiments among indigenous populations. An unknown author illustrated the confusion that natives felt on seeing mounted Spaniards for the first time: "[t]he Indians thought that the horse and its rider was all one animal, for they had never seen horses up to this time" (Schwartz 2000, 39). On conversing with a Nahua Indian, Diego Durán recalled the following prophesy that Indians correlated to the arrival of the Spanish: the vision told not only of the coming of the Spanish but that they were to arrive "mounted on beasts similar to deer and others on eagles which would fly like the wind" (Durán in Schwartz 2000, 39). Many eyewitness accounts bespeak the acute fear that the cavalier on his valiant steed induced, and visual representations from Native Americans recall this also. An armband crafted by indigenous Peruvians highlights the continued confusion and curiosity to which these new creatures gave rise. Art historian Cécile Fromont argues that the armband represents a space of correlation, whereby "the horse appears with the camelid feet of a llama rather than equid hooves, hovering in nature and status between the two worlds its original viewers were learning to straddle" (2014, 18). A rider sits astride a "horse," with long reins, in one hand, and what could be a riding crop, in the other, a long mane and tail loosely hanging from the animal. Given the fact that the horse is depicted with camelid feet—not the hooves of a horse—it is likely that the artist had not seen the rider or his mount up-close, or even first-hand; while an eyewitness seeing a horse from a distance would likely see stirrups and reins, the creature's hooves would be less visible, giving little reason to think the horse would have hooves different from those of the camels, with which indigenous Peruvians were familiar. The armband, dating from the sixteenth century, demonstrates how natives sought to understand the changing world around them, within which both the Spanish conquistador and the horse had become central figures.

If the first sighting of mounted Spaniards filled Indians with bewilderment and consternation, contact with these strange centaur-like creatures only exacerbated the unique situation. In the early years following Spanish arrival in the Americas, their cavalry greatly facilitated European domination of these new lands and peoples. Indigenous warriors were inadequately equipped to battle on an equal footing with their mounted counterparts. In the initial phase of the conquest, Native Americans sought in vain to protect themselves by adapting their battle strategies to match their new enemies, and many indigenous peoples succumbed to the Spaniard and his mount (Schwartz 2000).[2]

Horses not only facilitated Spanish conquest but further enabled conquistadors to become colonizers in these new lands. With horses, the Spanish could maintain a level of control over newly suppressed Indian populations, their unmatched speed offering Spaniards a means to travel from one location to another with great speed. Once new Spanish settlements had been established in the Americas, horses adopted roles akin to those they played in Europe: primarily in cultivating the land and conveying messages to neighbouring peoples. As Crosby argues, it was due to the horse that colonial America could develop such a successful cattle industry (Crosby 2003). While the role of horses was much more obvious in the first phase of the conquest, the horse nevertheless helped Spanish conquistadors to consolidate their newfound power in the Americas.

Across the Atlantic, in sub-Saharan Africa, horses were already familiar to indigenous Africans. Horses complemented camels in the Islamic caravans that traversed the deserts, linking the trade regions in north Maghreb, and had been present for many centuries. While North Africans successfully bred horses, the environment was not conducive to natural reproduction across West Africa: numbers of the tsetse fly, fatal to horses, increased during the sub-Saharan rainy seasons and thus practically precluded equine breeding. Several West African states, however, did succeed in creating and maintaining a mounted army, such as Oyo in contemporary Nigeria (later renowned for its cavalry), which relied on imported horses from North Africa (see Law 1976, 120). The arrival of Portuguese, and later Spanish, merchants along the West African coast offered Africans another source from which to fill their equine quotas.

The state of Oyo increased in size and power during the seventeenth century, a fact Paul Lovejoy correlates to an increase in their slave exports. By strategically engaging in commercial transactions with European merchants, Oyo sold slaves to negotiate for horses from North Africa (Lovejoy 2000, 81). Just as Spanish conquistadors were able to slaughter Native Americans, cavalry-states across West Africa held the upper hand in warfare. Building on scholars' studies of the "gun-slave-cycle," Robert Sydney Smith coined this process the "slave-horse cycle," whereby certain African nations became embroiled in a cyclical system of employing horses to capture slaves, which would then be traded or sold to European slavers on the coast for more horses, resulting in a perpetual, vicious circle. As long as Africans were unable to successfully breed horses across West Africa, they would be compelled to rely on imported equines (Smith 1989, 31). Horses provided growing African states with unprecedented force and speed, drastically tipping the socio-demographic geography of West Africa.

Centring our analysis on horses reveals a similar process in the post-conquest Americas and illustrates how Native Americans also appropriated horses arriving from Europe to serve their own needs. Following the initial shock, indigenous peoples from across the Americas quickly incorporated horses into their social and cultural lives. Across the Americas, indigenous peoples came to possess horses by multiple means, sometimes through warfare, as we see in Africa, but also through trade and treaties with European colonists, as will be shown with the example of the Appomattox Indians, in the following. Alfred Crosby asserts that the adoption of horses transformed the nature of Indian warfare—both across North and South America—enabling Indians to conduct swift raids for persons and booty, echoing the slave-raiding states of West Africa, and resulting in larger numbers of prisoners of war. Furthermore, Crosby goes on to demonstrate how, in an ironic twist, the Chichimecs from contemporary Mexico employed the horse in the same way their Spanish adversaries had done in the early years of conquest; indeed, a mounted army enabled the Chichimecs to retain independence from Spanish control for several generations (Crosby 2003, 104). In North America, the horse wrought a similar transformation in the lives of Native Americans. Across the plains, numerous Indian groups successfully resisted European domination into the late nineteenth century (Slatta 1997, 35). The Sioux Indians were driven from the forests of the north, where they also introduced the horse into their communities, the result being that by the 1860s, they were the "undisputed masters of an immense territory" and "the richest and most prosperous" of the Plains Indians (Mooney 1991, 824).[3]

Across the slaveholding Americas, both enslaved and free Africans utilized horses to resist white domination and control. The most iconic example is that of Touissant Louverture, an enslaved carriage driver, who led Saint Domingue's enslaved population to freedom. Slaves employed as carriage drivers were more able to pick up information of happenings outside of the microcosm of the plantations; such slaves were well-placed to convey messages across plantations faster than was possible on foot. When we look at those sources that detail slave rebellions or proposed rebellions—for instance, court records, images, and interviews with former slaves—we often see the presence of carriage drivers in slave rebellions or proposed rebellions.[4] While scholars have been quick to comment on the role that geographical movement allowed slaves, such as carriage drivers, it is important to stress that horses were central to enabling such movement.

Similarly, the slaveholding class employed horses to subdue slave revolts and to frighten slaves into subservience. Horses were central to rounding up runaways. A rider could cover a much greater distance than a runaway slave on foot (Lambert 2015, 6). Similarly, whites in the antebellum South saddled up each night to patrol the countryside, looking for slaves without passes permitting them to leave plantations after hours. One former slave, Fannie Moore from North Carolina, recalled to interviewers of the Works Progress Administration—who conducted interviews with former slaves from across the U.S. South—how a slave caught by the "paddyrollers" or the patrol system "had to show de pass to dem," and "if you had no pass dey strip you an' beat you" (*North Carolina Narratives*, Vol. X, Part 2, 132).[5] This system of patrolling southern plantations continued after the abolition of slavery. As Winthrop Jordan comments, it ultimately became a "tradition of nightriding that was used after the war to deal with blacks (particularly male) who got too uppity. . . . All that was needed was a horse, often a gun, and a length of cowhide" (Jordan 1993, 59). Mounted on a horse, such whites would become the forerunners of the Ku Klux Klan, the white man on a horse in a matching white cape and hat. Here, horses were central to whites, instilling fear and exerting a level of power.

2. Horses as a Disputed Symbol of Power and Prestige

The horse has a rich history denoting power and prestige across Western Europe. In England and France, a king's knights, or "chevaliers," occupied high positions in their countries' respective social hierarchies. In a French-English language dictionary dating from 1729, the term "chevalier" translates into English as "knight." The term "gentleman" or "gentilhomme" appears in association with "knight bachelour," or a "Gentleman who holds the position between a Knight and a Squire" (Boyer 1729, n.d.). Both the English and French definitions for "knight" convey great prestige, and those occupying the position of knight or chevalier would have generally been of high social standing. A similar statement can be made for the Spanish equivalent "caballero."

The association between horses and nobility was exported to the Americas with the arrival of the Spanish conquistadors. Colonial sources from the Americas, notably reports left by European colonists, highlight the importance of the horse as a newfound symbol of power and elevated status among Indian societies across both North and South America, illustrating how natives similarly adopted the European meanings accompanying horses. Stephanie Wood estimates that among the petitions submitted by "caciques"—a term employed in Spanish to denote indigenous American leaders or chiefs—to the Spanish colonial authorities, the most frequent request was the permission to ride a horse. In 1592, a Nahua noble "don Gabriel Curi received a license to carry a sword, wear Spanish clothing, and ride a horse" (Wood 2003, 49). Across the Spanish Americas, members of the indigenous elite recognized the importance that horses bestowed on riders and sought to acquire steeds to better reflect their elevated social status. Thus, if Spanish conquistadors rode horses, native leaders themselves would adopt horses so as not to be viewed as inferior, demonstrating once more how the cultural and social connotations surrounding the horse travelled from the European to the Native American milieu. It also shows the ambivalence of the master-slave dialectics: on the one hand, subalterns and subjugated peoples simply tend to interiorize dominant codes (in this example the horse as an icon of power), but on the other, by mimicking their masters, they try to acquire privileges and enhance their position and status.

The Anglo-Atlantic World further underlines the elevated status of horse and rider, showing how colonists continued to adhere to and enforce European social customs. Commenting on seventeenth-century Virginian tobacco-planter society, T.H. Breen argues that the horse was an essential component of a planter's life and identity. He writes that "[w]ithout a horse, a planter felt despised, an object of ridicule" (1977, 249). On the island of Bermuda, horses carried a similar level of prestige. While authorities introduced large taxes on horses in the early eighteenth century, Michael Jarvis purports that such levies merely resulted in further promoting islanders' desires to own a steed, the extra costs greatly raising the social prestige of this creature (2010, 313). Sailing southwards from Bermuda, colonists on the larger sugar-producing slave island of Jamaica also equated horses with power and prestige. In the blistering heat, Jamaicans regarded walking as highly degrading: slaves walked, whereas whites moved about the island on horseback or in horse-drawn carriages. In the *Jamaica Planter's Guide*, published in 1823, the author was clear to stress the importance of horse ownership to any aspiring planter: he wrote "how necessary it is for every white man on an estate or property to be possessed of a horse in his own right, in order to enable him, not only to discharge his public duty with ease, but to give him occasional recreation, with accustomed respect" (Roughley 1823, 68–9).

In the Anglo-Atlantic World, the connotations associated with horses passed from white colonists to Native Americans and black Africans. These peoples reacted very differently to contact with horses. Like their southern neighbours, North American Indians first reacted with bewilderment to the mounted European on horses. A letter dated 22nd August 1674 describes the "adoration" exhibited by the Appomattox Indians of contemporary North Carolina:

> a stake was sett up in ye middle of ye towne to fasten ye horse to, and aboundance of corne and all manner of pulse with fish, flesh and beares oyle for ye horse to feed upon and a scaffold sett up before day for my two men and Appomattocke Indian that theire people might stand and gaze at them and not offend them by theire throng.
>
> (Wood 2015)

The Appomattox treated the horses with almost as much indulgence as their English visitors. This initial bewilderment soon shifted as Indians across North America quickly began to see the advantages that such creatures bestowed on their owners. Indeed, decrees emerged prohibiting the sale of horses to Indians or prohibiting the sale of males, so that Indians could not breed horses (DeJohn Anderson 2006, 201). Colonists feared Indians may adopt horses into their military strategies: as long as only the English had a supply of horses, they continued to hold the upper hand—as had Spanish conquistadors—when pitted against Indians in warfare. Nonetheless, North Americans soon came to incorporate horses into their societies, despite colonial reservations. In the Plains, Indians adopted horses into their day-to-day routines, like in hunting game and gambling, to pay Shamans, and also in the most eminent of celebrations like the arrangement of marriages. As a consequence, the horse quickly "became the symbol of wealth and social status" (Slatta 1997, 47). In such ways, Plains Indians first adopted European symbolic notions of horses, re-appropriating these to fit with indigenous people's own respective cultural practices and values.

Looking to American slave societies offers another case study with which to see a similar appropriation of the cultural significance and symbolism of the horse. Across slaveholding societies, white masters brought with them perceptions of horses as symbols of prestige, as foregrounded in the examples of the various types of tasks owners assigned to favoured slaves. In the seventeenth century, South Carolina began importing slaves from Africa and the Caribbean. One slave profession was that of cattle herding, an occupation that bestowed a great amount of liberty on the slave caring for the livestock, given that he would often be away from his master for substantial amounts of time (Wood 1974, 19). For successful cattle herding, the slave required a horse. The master of the slave, horse, and the herd of cattle would have to hold the slave in high esteem and be sure of his loyalty and honesty to assign him such a prestigious role. Another task that conferred acute prestige was that of a carriage driver. A former slave from Indiana, Archie Koritz, who worked as a field labourer prior to his emancipation, recounted the following description of carriage drivers in the antebellum South:

> If a young, good-looking, husky negro was *trustworthy*, the family would make him the driver of the family carriage. They would dress him in the *best* clothes obtainable and with a silk-finished beaver skin hat.
>
> (*Indiana Narratives*, Vol. V, 203, my emphasis)

Another former slave from Arkansas remembered how his father had worked as a carriage driver under slavery. His final statement "[c]ouldn't trust 'em [the horses] with just anybody" underlines the importance of trust and honesty as characteristics of those slaves who undertook this profession (*Arkansas Narratives*, Vol. II, Part 5, 244). To become a carriage driver, a slave needed to display to his masters that he was of a trustworthy nature. If successful, he would receive fine attire topped with a hat—both symbols of stature for a person of slave status—and he worked in close contact with white company. Drivers often *heard* the gossip of their white masters as they drove the carriages. Andy Marion remembered how he "used to hear lots of things from behin' me, while drivin' de folks and saying nothing" (*South Carolina Narratives*, Vol. XIV, Part 3, 168).

Carriage drivers possessed a much greater level of geographical freedom than other slaves who were confined to the plantation. Their increased mobility gave them knowledge of life outside the plantations and opportunities for contact with freed blacks. Writing of Gabriel Prosser's slave rebellion in Virginia, James Sidbury argues that Woolfolk—a slave who was assigned the role of recruiting slaves from other plantations—did so on horseback because he exuded "power and authority" and would hopefully "impress those he wanted to recruit by arriving on horseback" (Sidbury 1997, 65). Following the 1812 Aponte Rebellion in Cuba, colonial authorities arrested and tried a number of black drivers (Childs 2006). The combination of geographical mobility, extended social connections, and access to information on whites' conversations frequently placed such slaves at the centre of planned rebellions and uprisings, the most notorious being that of Touissant Louverture.

Like Europeans, Africans also held the horse in high esteem. Travellers often commented on the close correlation between horses and the ruling elite or those members among the uppermost echelons of society, where horses occupied roles of prestige rather than merely as adjuncts in warfare.[6] In the kingdom of Dahomey, G. Robertson recorded in the early nineteenth century how "[t]hey have some horses, but they are not used except in processions, when they are ridden by the chiefs" (Robertson

1891, 265). Speaking of an encounter between a Congolese king and Portuguese ambassadors, John Leyden recounted how the king sported "on his shoulder a horse's tail, accounted here a peculiar ensign of royalty" (1817, 70). Another traveller to the Gambia furthered the importance of horses as symbols of power, writing that the "Great Master of the King's Horse" had in his control and possession an entire court (Moore 1738, 44). In Oyo, "horses were frequently kept as a tangible token of status by the rich and powerful: indeed, since horses were scarcer in Oyo, their value as a status symbol was perhaps greater" (Law 1975, 1). Given that the West African terrain did not favour the natural reproduction of horses, and as mentioned earlier, horses needed to be imported from the Maghreb North or from along the West African coast, facilitated by the arrival of European slave traders. Like Native Americans, Africans also saw how horses denoted power and prestige to their respective riders, and the eyewitness accounts speak to this adoption and appropriation of foreign cultural and social practices and values. As with any rare commodity, Africans paid or traded dearly for horses. Once in the Americas, it is not difficult to suppose that African slaves brought with them their own interpretations and understandings of horses.

A final interpretation of horses as symbols of power comes to light when one looks at the theme of sexual relations, and the frequent references to the dominant male figure as "riding" a woman, or physically mounting a female (or male) in an act of sexual gratification. In early Occitan literature, we see the analogy of a woman as a "horse" to be ridden by a male counterpart:

> I have two horses to my saddle, right and properly; good they are and skilled in war, and valiant. But I cannot keep them both, for one can't abide the other.
>
> If I could break them in to my desire, I would never wish to change my gear elsewhere, for I'd be better mounted than my living man.
>
> (William IX in Press 1971, 13)

The writer laments at how he finds himself in the unfortunate position of courting two women and his struggles at deciding with which woman to continue a sexual relationship, because the two women "can't abide" each other. The metaphorical representation of mounting and riding a woman and the author's use of the word "saddle" all underline the analogy between an act of sexual domination and the riding of a horse.

Fears that black men could have sexual domination over white women emerge when we look at slavery in North America. Records from a suspected slave uprising in Mississippi in the mid-nineteenth century underline the analogy between rape and the "riding" of a woman. When whites conducted interviews with slaves involved in the rumoured uprising, what particularly frightened them were the repeated references to "riding" the white mistresses and daughters. A testimony taken from the slave Harry Scott recorded that he intended to "kill old master and ride the ladies" (Jordan 1993, 284). The slaves would then "go on to Mrs O. Metcalfe and Mrs Dunbars and ride the ladies" (Jordan 1993, 273). Scott also claimed that the fellow slave Simon would also "be damned if he don't have one [a white woman] too" (Jordan 1993, 169). Jordan argues that for Simon, "riding the ladies *was* freedom. It was a fuller freedom because it was wanting by white women. . . . White women would want this because they would be running to really free men, and free men upheld their women" (Jordan 1993, 170). The slaves involved in the conspiracy demonstrated through their statements that they comprehended the correlation between freedom and the role of masculinity in southern society. For these men, freedom signified both control and protection of women, something that they were not able to fully fulfil in their enslaved status, and their testimonies speak to slaves' appropriation of white social practices. Moreover, such examples foreground the notion of a dominant male counterpart, a role that enslavement often sought to overshadow.

3. Horses as Agents *and* Symbols of the Magical and Spiritual Worlds

When we look at the religious representation of horses, we see with great lucidity similarities among former slave cultures in countries that maintain a strong historical link with African-rooted religions and religious practices. The religions of Candomblé and Voodoo both include a focus on spiritual possession: worshippers believe that a deity will come from the spiritual world and inhabit a particular devotee. This practice is commonly referred to as "riding," a term that foregrounds the metaphorical image of somebody riding a horse. Speaking of contemporary Haitian practices, Yvonne Daniel writes how "[t]he procedures train the body, the 'horse,' to receive Lwa energy." As Haitian Voodoo worshippers say, the procedures prepare their bodies to "gallop," first with an "empty saddle," then with a lwa "rider," and then to "obey his/ her commands" (Daniel 2005, 151). In Haiti, this practice is referred to as "monte chwal le," which phonetically translates as "monte le cheval" or "riding the horse" (Daniel 2005, 10).[7]

The imagery of "riding" likely has African roots. Across many African societies, metaphorical speech is ever-present and very important. In Yoruba, a famous saying states that "proverbs are the horses of speech" (Otero 2007, 48). Across the Atlantic in Brazil, those individuals who are in the state of being possessed by the "vodun" are referred to as "erê" or "esin," which has its origin in the Yoruba term "esin orisa," or "the god's horse," according to Luis Parés (Parés 2013, 258). The rider or deity, more commonly known by the West African terms "loas," "orixa," or "orisha," varies depending on the religious celebration, the personality of the person being ridden, and the god the practitioners have called. In all these instances, the joint imagery of the horse and the action of being ridden are central components to these spiritual ceremonies.

The Works Progress Administration interviews from the late 1930s with former North American slaves are a valuable source in this regard. An ex-slave from Georgia recounted that "when people die angry with someone they usually come back after death in the form of a witch and then they ride the person that they were angry with at the time of their death" (*Georgia Narratives*, Vol. IV, Part 4, 266); moreover, that witches or hags were "some kind of sperrits [spirits] an' dey would ride anybody" (*Georgia Narratives*, Vol. IV, Part 4, 266). Looking to earlier European folklore and popular culture offers a window into the origins of such beliefs that later permeated Afro-American cultures. The belief of witches or hags "riding" people during their night-time visitations was deeply rooted in European folklore. Henry Fuseli's oil painting *The Nightmare* illustrates this notion. The painting depicts the dark scene in which a woman lies prostrate on a bed, while a grotesque witch or hag mounts the dame's abdomen (Feingold 1982, 49). Overlooking this macabre sight is a horse, its eyes two white globes, its face inclined towards the woman's reproductive parts. Writing several decades after Fuseli's painting, Walter Scott observed the following with respect to the process of nightmares, mirroring Fuseli's visual representation: "In the nightmare an oppression and suffocation is felt, and our fancy instantly conjures up a spectre to lie on our bosom" (Scott 1830, 25). Similarly, in a verse entitled "A Nocturnal Sketch," an author writes of how the nursemaid lies "chest-press'd" while within the grip of a nightmare (Man 1833, 6).

The night-demon's frightening ability to constrict the circulation of the victim's heart also appears in the antebellum South in the nineteenth and twentieth centuries. In South Carolina, an ex-slave recounted how "[m]any of the old slaves, and some old white people" shared the belief that witches possessed the ability to ride people and, in so doing, were capable of stopping blood circulation (*South Carolina Narratives*, Vol. XIV, Part 3, 174). Penny Williams, formally enslaved in North Carolina, recalled an incident in which a witch attempted to "ride" her during the night: "one witch tried to ride me onct [*sic*]. I was in de bed, an' she thought dat I wus 'sleep. I feels her when she crawls up on my lef' leg an' stops de circulation" (*North Carolina Narratives*, Vol. XI, Part 2, 405).

The examples discussed previously speak once again to the ways in which a cultural perception of the horse could be appropriated by other social groups across both different geographical and temporal scales in the Atlantic World. Moreover, these analogies demonstrate the interrelation between the world of the living and the world of the spirits. Through the symbolism of the horse, these two worlds come together: the spiritual world can enter and influence the world of the living. Yet, unlike the examples from Brazil and the Caribbean, where participants appear to have actively sought to be "ridden" by gods, North American ex-slaves expressed that they did not seek to be ridden or possessed by these otherworldly spirits, which typically happened when the person was asleep. Although the interviewees did not refer to the source of the term "riding," the fact that the verb appears in a very similar context in other former slave societies across the Americas would suggest that the term did indeed have shared roots, presumably from Africa. Given that these sources come from former slaves and descendants of slaves from slave societies in the twentieth century (and into the twenty-first century in some instances), we see a continuation over both great temporal and geographical scales. It is reasonable to assume, therefore, that such beliefs had existed among slave and black communities during the era of African slavery.

4. Conclusion

Horses occupied a very central role in both the formation and the evolution of the Atlantic World. As this chapter has demonstrated, various peoples and social groups from across three continents—Africa, the Americas, and Europe—often shared similar views and practices of this equine beast, allowing us to build on our already-growing understanding of the interconnectedness of the various peoples of the Atlantic World. I have argued that the horse as a case study allows us to identify the existence of similarities among historical actors who, at a first glance, perhaps appear rather distant culturally or socially from one another. In addition, I have demonstrated how such similarities in the cultural perceptions of horses across social groups laid the foundations for future beliefs which were appropriated and re-appropriated. The case of the horse illustrates how beliefs, customs, ideas, and values migrated across the Atlantic World and how such variables shifted in meanings across different social groups.

We have seen the importance of horses in three predominant areas: first, that of agents or actors of power; second, as symbols of prestige and power; and finally, as symbols of the spiritual sphere. The overarching theme that continues to emerge throughout each of these three sub-contexts is that of power. Whether a slave carriage driver on a North American cotton plantation, a Native American chief riding into battle, a Spanish conquistador astride his armoured steed, Toussaint Louverture galloping to victory and freedom in Haiti, or the spiritual "ridings" of African-rooted religions, these examples underline the multiple ways whereby the horse offered its riders across the Atlantic World an opportunity at an authority, power, and elevated esteem, which he might not have the same access to without a mount. This is Atlantic history from the saddle.

Notes

1. See Melville (1997), Crosby (2003), DeJohn Anderson (2006), and Few and Tortorici (2013).

2. As Schwartz observes, when confronted by the Spanish cavalier, the Mexica "sought terrain that neutralized the advantage the Spaniards gained with horses" (2000, 12).
3. In this example, Native Americans successfully evaded European colonization by appropriating the horse as an agent of power and tool of warfare. See also Thornton (1987).
4. See for instance Sidbury (1997) and Childs (2006).
5. Included in the collection "Born in Slavery: Slave Narratives from the Federal Writers' Project, 1936–1938," http://memory.loc.gov/ammem/snhtml/snhome.html (last accessed 3/23/2001). Subsequent citations from this collection will be referenced as follows: *North Carolina Narratives*, *Arkansas Narratives*, etc.
6. While this chapter has demonstrated that horses and armed cavalry were central to the expansion of numerous African states geographically and politically, horses did not occupy a solely military role. In her controversial article on the horse in fifteenth-century Senegambia, Elbl argues that horses were much more likely to denote power and prestige than to be employed in warfare (1991, 94).
7. Author's translation.

References

Arkansas Narratives, Vol. 2, Part 5.

Armitage, David (2009). "Three Concepts of Atlantic History." In *The British Atlantic World, 1500–1800*, edited by David Armitage and Michael J. Braddick. New York: Palgrave Macmillan.

Bailyn, Bernard (2005). *Atlantic History: Concept and Contours*. Cambridge: Harvard University Press.

Benjamin, Thomas, Hall, Timothy, and Rutherford, David (eds.) (2001). *The Atlantic World in the Age of Empire*. Belmont: Wadsworth.

Boyer, Abel (1729). *The Royal Dictionary, French and English, and English and French*. London: J. and J. Knapton et al.

Breen, T.H. (1977). "Horses and Gentlemen: The Cultural Significance of Gambling among the Gentry of Virginia." *The William and Mary Quarterly* 34, no. 2: 239–59.

Catlin, George (1866). *Illustrations of the Manners, Customs, and Condition of the North American Indians With Letters and Notes*. Vol. 2. London: Henry G. Bohn.

Childs, Matt D. (2006). *The 1812 Aponte Rebellion in Cuba and the Struggle against Atlantic Slavery*. Chapel Hill: University of North Carolina Press.

Crosby, Alfred (2003). *The Columbian Exchange: Biological and Cultural Consequences of 1492*. 30th Anniversary Edition. London: Praeger.

Daniel, Yvonne (2005). *Dancing Wisdom, Embodied Knowledge in Haitian Vodou, Cuban Yoruba, and Bahian Candomble*. Urbana: University of Illinois Press.

DeJohn Anderson, Virginia (2006). *Creatures of Empire: How Domestic Animals Transformed Early America*. Oxford: Oxford University Press.

Elbl, Ivana (1991). "The Horse in Fifteenth-Century Senegambia." *The International Journal of African Historical Studies* 24, no. 1: 85–110.

Feingold, Lawrence (1982). "Fuseli, Another Nightmare: The Night-Hag Visiting Lapland Witches." *Metropolitan Museum Journal* 17: 49–61.

Few, Martha, and Tortorici, Zeb (eds.) (2013). *Centering Animals in Latin American History*. Durham: Duke University Press.

Fromont, Cécile (2014). *The Art of Conversion: Christian Visual Culture in the Kingdom of Kongo*. Chapel Hill: University of North Carolina Press.
Georgia Narratives, Vol. 4, Part 4.
Indiana Narratives, Vol. 5.
Jarvis, Michael J. (2010). *In the Eye of All Trade: Bermuda, Bermudians, and the Maritime Atlantic World, 1680–1783*. Chapel Hill: University of North Carolina Press.
Jordan, Winthrop D. (1993). *Tumult and Silence at Second Creek: An Inquiry Into a Civil War Slave Conspiracy*. Baton Rouge: Louisiana State University Press.
Lambert, David (2015). "Master-Horse-Slave: Mobility, Race and Power in the British West Indies, c.1780–1838." *Slavery & Abolition* 36, no. 4: 618–41.
Law, Robin (1975). "A West African Cavalry State: The Kingdom of Oyo." *The Journal of African History* 16, no. 1: 1–15.
Law, Robin (1976). "Horses, Firearms, and Political Power in Pre-Colonial West Africa." *Past & Present*, no. 72: 112–32.
Leyden, John (1817). *Historical Account of Discoveries and Travels in Africa, Enlarged and Completed to the Present Time*. Vol. 1. Edinburgh: George Ramsay and Company.
Lovejoy, Paul E. (2000). *Transformations in Slavery: A History of Slavery in Africa*. 2nd Edition. Cambridge: Cambridge University Press.
Man, Thomas (1833). *Picture of a Factory Village: To Which Are Annexed, Remarks on Lotteries*. Providence: Printed for the Author.
Melville, Elinor G.K. (1997). *A Plague of Sheep: Environmental Consequences of the Conquest of Mexico*. Cambridge: Cambridge University Press.
Mooney, James (1991). *The Ghost-Dance Religion and the Sioux Outbreak of 1890*. Lincoln: The University of Nebraska Press.
Moore, Francis (1738). *Travels Into the Inland Parts of Africa*. London: Edward Cave.
North Carolina Narratives, Vol. 11, Part 2.
North Carolina Narratives, Vol. 11, Part 2, Interview with Fannie Moore, ex-slave.
Otero, Solimar (2007). "Spirit Possession, Havana, and the Night: Listening and Ritual in Cuban Fiction." *Western Folklore* 66, no. ½: 45–73.
Parés, Luis Nicolau (2013). *The Formation of Candomble: Vodun History and Ritual in Brazil*. Translated by Richard Vernon. Chapel Hill: University of North Carolina Press.
Robertson, G.A. (1891). *Notes on Africa: Particularly Those Parts Which Are Situated Between Cape Verd and the River Congo*. London: Sherwood.
Roughley, Thomas (1823). *The Jamaica Planter's Guide: Or, a System for Planting and Managing a Sugar Estate, or Other Plantations in the Island*. London: Longman et al.
Schwartz, Stuart (2000). *Victors and Vanquished: Spanish and Nahua Views of the Conquest of Mexico*. Boston/New York: Bedford-St. Martin's.
Scott, Walter (1830). *Demonology and Witchcraft, by Sir Walter Scott*. London: John Murray.
Sidbury, James (1997). *Ploughshares Into Swords: Race, Rebellion, and Identity in Gabriel's Virginia, 1730–1810*. Cambridge: Cambridge University Press.
Slatta, Richard (1997). *Comparing Cowboys and Frontiers: New Perspectives on the Histories of the Americas*. Norman: University of Oklahoma Press.

Smith, Robert Sydney (1989). *Warfare & Diplomacy in Pre-Colonial West Africa*. Madison: University of Wisconsin Press.

South Carolina Narratives, Vol. 14, Part 3.

Thornton, Russell (1987). *American Indian Holocaust and Survival: A Population History Since 1492*. Norman: University of Oklahoma Press.

William IX (1971). "William IX, Count of Poiton, 'Duke of Aquitaine.'" In *Anthology of Troubadour Lyric Poetry*, edited by Alan R. Press. Edinburgh: Edinburgh University Press.

Wood, Abraham (1674). "The Journeys of James Needham and Gabriel Arthur in 1673 and 1674 through the Piedmont and Mountains of North Carolina to Establish Trade with the Cherokee." *The Cherokee Heritage Documentation Center Website*. Accessed December 6, 2015. http://cherokeeregistry.com/index.php?option=com_content&view=article&id=406&Itemid=615.

Wood, Peter H. (1974). *Black Majority: Negroes in Colonial South Carolina From 1670 through the Stono Rebellion*. New York: Alfred A. Knopf.

Wood, Stephanie (2003). *Transcending Conquest: Nahua Views of Spanish Colonial Mexico*. Norman: University of Oklahoma Press.

4 Re-animalizing Animals, Re-animating Humans[1]

Michał Krzykawski

In his book on how the act of seeing establishes the human place in the world, John Berger writes that "the way we see things is affected by what we know or what we believe" (2008, 8). However, as he continues, "we only see what we look at" since "to look is an act of choice" (8). To look at someone—this animal or that wo/man—is always to look after them, and all those cats and dogs we live with might serve as a potent case in point here. Our act of choice is precisely to pay attention to our pets but to turn our gaze away from those animals we feed them with or eat ourselves. A quick look at my cats' stockpile full of "natural cat food," which strangely resembles regular middle-class dishes, might be an example: chicken breast with wild rice, pumpkin, or asparagus, trout with a julienne of carrots and tomato, beef with carrots, tomato, and courgettes, and coley with spinach. It is fairly enough to make my cats look like those that get the cream whenever five 'o clock hits. Besides, they always try to convince me that the clock regulating the human *Umwelt* is slow. One of them jumps on my desk, bumps her head against mine, and walks on my laptop back and forth from half-past three, while the other hangs around or takes her seat on the floor and looks up at me in a wait-and-see attitude. It is our daily routine.

I am not saying this because I am a cat lover. The cats I live with, to be cold-bloodedly fair, are not important in this pet-ty story about marketing stupidity that not only forms and formats the human imagination but also it affects our pets. I am rather interested in those singular animal existences—now packed and canned—that I could not see before they were turned into "natural cat food." In fact, my inability to look at them as they were is not an act of choice. The daily routine is then accompanied by the same question that reappears every day more-or-less persistently: do I make my cats "eat well"? (Derrida 1991, 115).

This chapter deals with the industrial process of killing animals. The fact that we humans feed our *anthropomorphized* pets, as full members of human-animal Western families, with *processed* and *de-animalized* animals is one of the side-effects of this process. I should like to locate my argument on the margins of Florence Burgat's last book, entitled

L'humanité carnivore (2017). My attempt is not to conduct "a moral critique developed by promoters of a vegetarian lifestyle" because I believe, in the wake of Burgat's reasoning, that "those promoters provide us with no conceptual tools to understand why humanity is carnivorous" (2017, 32–3).[2] However, unlike Burgat, whose attempt is to challenge "the carnivorous fact" in relation to "the initial fiction" that established "humanity as a *whole*" and resulted in "metaphysical and moral implications soon accompanied by political power" (16–17), I will only focus on the human-global carnivorousness as a disruptive innovation[3] of the capitalism of the second half of the twentieth century, putting an emphasis on how this carnivorousness has affected the way we feed our humanized pets and how their acculturation (rather than domestication) makes animal ethics extremely vulnerable.

Thus, moving beyond the ethical approach to animals, the first part of this chapter offers a cross-reading of Karl Marx's "general intellect" (1973) and Éric Baratay's "animals-machines" (2012, 261) in the context of what Bernard Stiegler describes as the "hyper-industrial epoch" (2014b, 2015). Animals are definitely not Stiegler's concern. Moreover, when admirably describing the incompleteness of what he refers to, in the wake of Aristotle, as the noetic soul, and why this soul needs to be constantly supplemented by the artificial forms of *technē*, Stiegler follows the Kantian conceptualization of animality as the origin above which man must rise in order to *form* their humanity (de Fontenay 1998, 727). In consequence, the noetic soul, as it is seen by Stiegler, contrary to the philo-logical *sense* of *noūs* (Cassin et al. 2014, 2555–9, has no "animalian foundation [*fonds*]" (Stiegler 2011, 136): "When man strives for the animal, they sink into bestiality at most times, while an animal *cannot* be bestial: an animal can be nothing but an animal imperturbably" (Stiegler 2005, 58). Therefore, thinking the animal with Stiegler means to go beyond his concept of the noetic soul. In fact, when Stiegler shows that "the industrial exploitation of the drives," which is at the heart of the consumerist capitalism, produces a systemic stupidity, and when he posits, at the same time, that "it is in principle possible to struggle (while under no illusions about the possibility of victory) against stupidity" (2015, 17 and 32), he somehow offers me a path to follow. However, he also encourages me to go further and claim that this struggle is impossible with no regard to animals who, as Derrida says, "cannot be stupid" (2009, 51). It is then possible to struggle (while under no illusions about the possibility of victory) *for* animals. If the reinvention of a viable future in the Anthropocene—which engages us and our *noūs* in the struggle against mindlessness and demands that we recover the political sense of *philia* (Stiegler 2003, 28)—is the motive of Stiegler's "new critique of political economy" (2010), this struggle also has to respond to *nos amis les bêtes* due to the simple fact that animals (*les bêtes*) challenge our exclusive mindlessness (*bêtise*), which we erroneously identify with them for our greater convenience.

For I believe, following Stiegler, that "logical disputation[s] [are] required for the exercise of reason" and lay "at the origin of t[he] disciplines called rational" (2015, 181 and 27), and I argue that the scene of this politico-economical struggle is first and foremost philosophical. In the second part of this chapter, my goal is to confront Stiegler's critique of political economy with Florence Burgat's animal phenomenology and, as a result, to modify his approach to animal life. Yet, notwithstanding my highest interest in Burgat's remarkable attempt to conceptually dissociate sacrifice (understood as making an offering) from the sacrificial structure (understood, in the wake of Derrida, as a place left open for a noncriminal putting to death) (Burgat 2017, 391), I will express some reservations about her thinking "with Freud and against Marx" (361) and, as a result, her underestimation of the role of technics (and technology) in relation to what I refer to as de-animalization. Thus, thinking *both with Freud and Marx*, and through a cross-reading of Burgat and Stiegler, I will attempt to push the animal question towards a critique of political economy.

1. Animals as Objects of a Critique of Political Economy

My investigation of the ways of seeing animals beyond ethics stems from a simple observation that animal lives are invisible in the hyper-industrial epoch, while the self-destructive power of consumerist capitalism affects animals and their places to live in the most unimaginable ways. Crudely speaking, no ethics is actually possible in relation to those whom we do not see, even though we do know that they are (t)here. In fact, as "non-inhuman beings faced with the fact of being-inhuman" (Stiegler 2015, 47), we do know that those invisible animals, counted in billions, are sentient and intensely suffer as killable livestock in *hyper-industrial* animal industries. We know it, and yet we feel that *their actual catastrophe* does not belong to *our reality*, no matter what we do in our daily lives: whether we go vegan, demonstrate against hunting, or eat a steak. Though we still cannot imagine a world without animals, who have inhabited our imagination "since so long ago [*depuis le temps*]" (Derrida 2008, 2), we are unable to *see them as they are* and *see how they are* in this world.

What Stiegler refers to as the "hyper-industrial epoch" is, broadly speaking, modernity as the main feature of industrial society: "I refer to a society as 'hyper-industrial' when everything in it has become subject to modelling and industrial activity—distribution, health, leisure, education, and so on—a point of view that is obviously a complete break with the myth of 'post-industrial society'" (2015, 228). This hypothesis clearly challenges that of Lyotard from his famous "report on knowledge," which became, from its publication in 1979, a dominant theoretical reference of the so-called "post-industrial era" and "postmodern era," which societies and cultures entered, respectively (Lyotard 1984, 3). However, if "there is no 'post-industrial' future" (Stiegler 2015, 171), one may wonder whether there is a future for animals in hyper-industrial society. In fact, the relationship between humans and animals drastically changed with the advent of industrial society, whereas the constitution of sociology as a discipline radicalized the distinction between (animal) nature and (exclusively human) society (Despret and Porcher 2007, 14–16).

At the beginning of his study, "Why look at animals?," John Berger says:

> The 19th century, in western Europe and North America, saw the beginning of the process, today completed by 20th century corporate capitalism, by which every tradition which has previously mediated between man and nature was broken. Before this rupture, animals constituted the first circle of what surrounded man. . . . They were with man at the centre of his world. Such centrality was of course economic and productive. Whatever the changes in productive means and social organisation, men depended upon animals for food, work, transport, clothing.
>
> (Berger 2009, 1–2)

However, a significant transformation took place with this completion that I will discuss as a disruptive innovation within hyper-industrial and post-animal society. On the one hand, animals were not any more "animals-machines used for production purposes" (Baratay 2012, 261), and, as a result, they lost their objectified position. On the other hand, their *bodies* became *disposable* and *processable* by means of what Marx described as "organs of the human brain, created by the human hand." Indeed, in order to understand what disruptive innovation is, and why it stems from an unprecedented transformation of the social status of animals, a quick look at *Grundrisse* is necessary. Marx says:

> Nature builds no machines, no locomotives, railways, electric telegraphs, self-acting mules etc. These are products of human industry; natural material transformed into organs of the human will over nature, or of human participation in nature. They are *organs of the human brain, created by the human hand;* the power of knowledge, objectified. The development of fixed capital indicates to what degree general social knowledge has become a *direct force of production,* and to what degree, hence, the conditions of the process of social life itself have come under the control of the general intellect and been transformed in accordance with it.
>
> (1973)

When one reads this excerpt from "the animal point of view," to quote the title of Éric Baratay's book (2012), the transformation of "animals-machines" into *disposable* and *processable material* can be read between the lines. Previously used and abused as "organs" in the process of "human participation in nature," animals are hence processed by those "organs" themselves; they become "natural material transformed into" *valuable goods*—not "organs"—by means of what Marx refers to as "general intellect," that is to say, technological expertise and social knowledge embodied in "fixed capital" (machines and infrastructures as means of production). In the terms used by Marx, animals are dismissed from the "position of instruments" and become "subjects of labour" (raw materials provided by nature de/formed and modified by man) (Marx 1887). The transition from industrial society to hyper-industrial society appears afterwards in the most violent way from the perspective of the hyper-industrial captivity of dissocialized animals, which will have become one of the least discussed issues of consumerist capitalism.

In his latest books, Stiegler describes consumerist capitalism and its calamitous impact on every form of social life with the term of disruption (2016, 2018). According to him, this term nowadays refers to "an epoch of innovation in which exosomatisation [the fact that human organs are primarily the technological organs we depend on, MK] is now completely controlled by economic powers and subject to the constraints of short-term profitability" (Stiegler 2018, 90). Nevertheless, in the field of business, disruptive innovation—described in Harvard Business Review—referred to a technological innovation which launches a new market and value network and makes an existing market and value network fail (Bower and Christensen 1995). Then, one may wonder whether the de-animalization, i.e. the transformation of animals into disposable and processable material by means of "general intellect," might be considered *the most disruptive innovation* in the history of hyper-industrial society since it broke the human-animal relation and, as a result, made meat consumption explode in the 1950s (Baratay 2012, 295), radically changing our dietary habits and requirements on a *global* scale. Indeed, as Florence Burgat points out,

> even countries in which meat is not the basic element of traditional cuisine are pushed to come into the era of ovo-lacto-meaty plethora [*pléthore ovo-lacto-carnée*] by industrial superpowers which work to generalise the Western lifestyle and alimentary model which, consequently, work to profit them.
>
> (2017, 9)

"The general intellect" is a general *condition* of the animal lives in the industrial world. Curiously enough, animals as "instruments" could count on more consideration by the men who worked *with* them than the animals totally subjugated to and deformed by human machines.

Describing "industrial horses" (Baratay 2012, 107) working in French coal mines at the beginning of the twentieth century, Baratay convincingly shows that a relationship was possible between *this or that singular* animal and his owner/foreman who looked after his machine. This form of attention encouraged the animal-machine not to resist and to work more efficiently (119–134). Animals *processed* by "organs of the human brain," in turn, must have been de-objectified to become nothing but *hyperindustrial material*. As "subjects of labour," they were not seen as the objects (and potential subjects) of ethics. The invisibility of animals *as such* became even more *obvious* with the advent of "cognitive capitalism," which Moulier Boutang describes as a "cooperative labour of human brains joined together by networks" or "a cooperation of brains in the production of the living by means of the living" (2011, 57).

Cognitive capitalism somehow sealed animals' fate in hyper-industrial society by eliminating them from the (exclusively human) living which (re)produces itself and introduces new forms of human alienation in a new post-animal world. This transformation of the human perception of animals was the next stage of the process analysed by Baratay from a historical perspective: if industrial animals, such as horses or cows, whose singular lives are recreated by Baratay, could have been seen as *lumpenproletariat* at the beginning of the twentieth century, they have *massively* become "quickly used and replaced objects of consumption" (Baratay 2012, 169) since the 1950s. Post-industrial society is a myth, as Stiegler claims, also because this society, from its very mythical beginning, is grounded on the hyper-industrial exploitation of animals, whereas the general intellect makes this exploitation more and more invisible and doomed to oblivion. This is precisely why animals, I argue, have never been the objects of any critique of the political economy, Adorno's and Horkheimer's effort notwithstanding. Moreover, related to what the two philosophers described as "the intertwinement of rationality and social reality, as well as of the intertwinement, inseparable from the former, of nature and the mastery of nature" (2002, xviii), the exclusion of the animal from the living significantly challenges the validity of animal rights. In fact, the elusive animal rights were a dead letter even before they became the object of a serious social debate in hyper-industrial society. The industrial animal is hence considered the living with no right to live: *actually* the non-living.

Understanding this transformation of human perception in relation to animals is of crucial importance here. On the one hand, it lets us see the *actual* animal presence in a modernity shaped both by human machines and by "animals-machines" in a much more material way than what Giorgio Agamben refers to as "the anthropological machine of humanism" (2004, 28). On the other hand, it shows us that the advent of hyper-industrial society coincided with the process of throwing animals out of "social life . . . under the control of the general intellect" and was the first step towards the *post-animal*—rather than posthuman or transhuman—world. Pragmatically speaking, animal invisibility does not (only) result from the human anthropocentrism which, "until the 19th century, [. . .] was integral to the relationship between man and animal and was an expression of their proximity" (Berger 2009, 9) Animal invisibility is, rather, a *catastrophic* effect of de-animalization which preceded and made *systematically* possible the explosion of meat consumption in the 1950s (Baratay 2012, 295), which was tantamount to the animal *catastrophē*, that is to say, according to the ancient Greek, their overturn or solution. At this point my reasoning follows Jes Lynning Harfeld, who claims that "the invisibility of the animalness of the animal constitutes a fundamental obstacle to change within current production systems" (Harfeld et al. 2016).

The process of de-animalization turned animals into unanimated matter. Deprived of their essence, animals are hence less than *zoē*—their life is no more a "sentient life" that man, to put it in Aristotle's terms, shares with "the horse, the ox, and every animal" (2000, 12, 1098a), while the value of this non-*zoē* appears only when it has become *processed*. De-animalization seems to be the tragic final stage of the consecutive processes of the domestication and instrumentalization of animals in the modern era. If the instrumentalization of animals pulls them out of nature, the process of de-animalization not only denies them their utility in the human-centred world but also *systematically* refuses them their biological (and psychological, as I will show in the following) right to life as living organisms in the transformed world. The human-animal relation is *disrupted*. However, this disruptive innovation could have occurred because it stems from a long historical process initiated by Industrial Revolution: man loses sight of animals as *he* abandons the animal he has domesticated and forced to work.

When John Berger says, "[w]hatever the changes in productive means and social organisation, men depended upon animals for food, work, transport, clothing" (2009, 1–2), he seems not to notice this disruption being achieved within the general process he describes, a disruption which led us to a new *post-animal oikos*. Humans became economically independent from animals, whereas animals themselves became totally dependent upon humans. As Florence Burgat points out, animal butchery becomes generalized, industrialized, and standardized when mankind can do without killing in order to feed itself (2017, 10). By 1980, this disruption had perhaps not yet come to light, for it belonged to the world which was being driven into the new economic order of the post-animal world. The history of animals in (hyper-)industrial and post-animal society remains to be written, though Éric Baratay's works have already paved the way for it (2012, 2017). Such a history would make the point clearer that the rise of neoliberalism, marked by a dubious liberation of men from their milieu, was tantamount to the fall of animals and their de-animalization as the final result of what Horkheimer and Adorno described as a "corrosive rationality" (2002, 4).

2. Economics Is Before Ethics

The invention of wet and dry food intended for consumption by dogs and cats is perhaps one of the most salient, though the least discussed, features of de-animalization. Hence dogs and cats, whose significance in occidental families has been systematically growing since the end of the nineteenth century, live on a regular diet of de-animalized and processed animals. The latter's invisibility coincides with a selective anthropomorphism, which makes us *exclusively* look at our pets. Baratay points out that feeding dogs with meat in the 1950s was a consequence of the explosion of meat consumption by men and, simultaneously, the popularization of the idea of canine carnivorousness—even though dogs, until the end of nineteenth century, were fed on products of plant origin, reflecting man's contemporary alimentary practices (Baratay 286–295). In the 1960s, when food intended for consumption by dogs appeared on the market, their diet changed once more. However, as Baratay continues, dog food was initially made of *animal* materials considered waste and had an obvious impact on the health of dogs. Nutritionally valuable dog food has only been produced since the 1980s. At the same time, dog- and cat-owners have become the targeted audience for pet junk food TV commercials which reiterated images of happy and healthy dogs (or cats, which regretfully do not form the object of Baratay's historical analyses) as full and fully humanized members of the carnivore human-animal families. The power of those images stems from the extremely rapid growth of television, whose invisible counterpart was the extremely rapid disappearance of animals from view.

Faced with this invisibility, the ethical approach to animals seems to be in a somehow desperate quest for its *object*, while the economically binding model does not let us go further than to make the industrial process of killing animals more "humanitarian." A recent French law which demands slaughterhouses to install cameras in order to protect animals improperly killed may serve as an example of this *systemic* aberration. It does not even pretend to work to change the production system which makes animals invisible but rather makes the system even more technologically advanced and as trouble-free as possible, subordinating the process of de-animalizing animals under the control of the new instruments of "general intellect."

In a conversation with Jean-Luc Nancy, first published in English in 1988, Jacques Derrida points out that "the limit between the living and the non-living now seems to be as unsure, at least as an oppositional limit, as that between 'man' and 'animal'" (1991, 15). However, when taking into consideration the extent to which the explosion of meat consumption disrupted the social relation between humans and animals in hyper-industrial society and, no less importantly, the simultaneous change in human relationships with dogs and cats, one may argue that this boundary seems to be very clear, though it rather divides human-animal families from the industrial animal, i.e. the living from those lacking the right to live. Derrida's approach to the animal is first ethical. The question of eating animals, strictly related to the "appropriation-assimilation of the other" (15), comes back to that of how a subject becomes one. In fact, Derrida's famous injunction *il faut bien manger*, which is to be understood both as one has to eat *well* and one needs to eat *definitely*, is nothing but a variation of his constant inquiry: how to be a proper subject, which can also be taken as "clean" and "white" (*proper*) subject, and how to "determin[e] the best, most respectful, most grateful, and also most giving way of relating to the other and of relating the other to the self," since the other, which is always already (t)here, challenges the property of myself (15). One may ask, however, whether seeing the animal—who looks at me and is my concern (*me regarde*), as Derrida famously puts it when following his "autobiographical animal" (2008)—as the other might be a helpful tool in relation to industrial animals, since the hyper-industrial conditions dis-figure their otherness quite radically, i.e. deprive them of their face [*figure*]. Although the concept of the other connected many French thinkers of the second half of the twentieth century, it seems to be too abstract to respond both to the non-otherness of industrial animals and our anthropomorphized pets, these industrial meat-eaters we care for as if they were like us.

My objection to Derrida's approach to the animal as the other is not meant to reject deconstruction in any way. I rather assume, with a nod to Catherine Malabou, that deconstruction has taken place (*a eu lieu*) (2010, 52), which is both comfortable and demanding: if the deconstructive dismissal (*non-lieu*) of the proper of men—which calls for justice and responsibility—can be henceforth taken for granted at least in our academic debates, the very heritage of deconstruction requires us to go further, be it with or against Jacques Derrida. In a similar vein, i.e. through the animal *default* of Stiegler's philosophy of *technē*, I rather try to see how the animal question has been predetermined by technics (and technologies) in hyper-industrial society and how animal life has been put at the service of the free market economy. Challenging Lévinas' conviction according to which "ethics is before ontology" (1982, 143), I would argue that, as far as animals are concerned, economics is before ethics. Through the disruptive innovation which made of us ovo-lacto-meat-eater consumers on a global scale, economics has radically displaced animal ethics by displacing animals *as such* from society, or from what Jacques Rancière refers to as "the sensible" (2004), i.e. primarily *the visible*. The real physical disappearance of animals from our field of view, and their *unreal* reappearance in multiple productions of "cultural industry," has somehow annulled ethics at its very origins.

Of course, Jacques Derrida was aware of this displacement. In an interview with Élisabeth Roudinesco, originally published in 2001, when talking about the animal industry, he said

> The relations between humans and animals *must* change. They *must* both in the sense of an "ontological" necessity and of an "ethical". I place these words in quotation marks because this change will have to affect the very sense and value of these concepts (the ontological and the ethical).
>
> (2004, 64)

And yet, no change is possible until animals have been re-animalized by humans.

New *concepts* are needed to think (and look after) animals. The invention of such *technical* tools—far beyond post- and transhumanism, which do not seem to offer much in this respect, as these academic trends rather seem to bring down the curtain on economic issues—is one of the most urgent duties of contemporary philosophy in its irrevocable relation to the critique of political economy. In fact, to recognize animals in their actual existence or—to put it more precisely—as actual existences means to extract the animal question from the post- and transhumanist discourse. It is a matter of urgency to think the plural animal existences without hammering away at man's anthropocentrism, i.e. to approach animals *reasonably* (de Fontenay 2008) by means of what Malabou, in a Kantian critical vein, defined as "rational imagination" (2010, 13), and opt for what I would call *humanimalities*, which might animate and re-animate the humanities in general. Drawing on the digital humanities programmed by Stiegler, the humanimalities, whose main concern would be human-animal life in relation to *technē*, could also struggle hand in hand against "the atrophy of the life of the spirit" and "epistemological error" (Stiegler 2014a, 15–16), which has erased human-animal relations.

3. Clean Meat as a Positive Disruptive Innovation?

In her animal phenomenology, Florence Burgat consequently tries to eliminate this epistemological error, although her motive is totally different from Stiegler's. In its most rudimentary sense of trying to see objects as they are originally given to consciousness, phenomenology becomes a potent tool to see animals afresh *as they appear*, that is to say, to leave behind all assumptions about animals. Crudely speaking, before any attempt to see animals as singular phenomena appearing as they are, one must *suspend* all he or she "knows" about them. That this "phenomenological reduction," as Husserl has it, is particularly difficult in relation to animals proves how cultural beliefs and psychological conditions, which overdetermine the institutionalized "knowledge" about animals, make us unable to see them, *our best intentions and animal-friendly attitudes notwithstanding*. What is challenged here is perhaps not human anthropocentrism as a decisive reason for the human blind spot but rather human exceptionalism (not to be confused with human *singularity*). Quite likely, suspending all the assumptions we have about animals, ground-breaking assets of ethology notwithstanding (Despret 2012), is impossible if not undesirable, since desire, with all affects it calls upon, makes us want to know them by means of understanding: both sympathy and intellect (with a nod to what Kant defined as *Verstand*). However, phenomenology remains a helpful tool, as it enables us to see animals as objects and consequently to give them back their actual (and plural) animalness with respect to human (and plural) singularities.

Burgat's approach to animals is triple. Firstly, drawing on Hegel's observation that "the animal cannot escape a feeling of *insecurity*, *anxiety*, and *misery*" (Hegel 2013, 179), Burgat posits "the *non-naturality* of the animal life" (2012, 15). *An* animal is a *subject* because it *feels*, that is to say, is *pushed* to escape the world's hostility in order to *live*. The way Burgat benefits from Hegel's observation is of the highest interest. On the one hand, she extirpates animals from the human sphere, i.e. from the notion of animality as a reverse side of the human. On the other hand, she shows that animals, unlike vegetative forms of life, do not belong to the sphere of nature, neither do they depend on the laws of nature. In a nutshell, animals have inner lives. In the foreword to Burgat's *Une autre existence*, Heinz Wismann suggests the potent political implications of Burgat's approach: her animal phenomenology makes it possible to consider the animal life as *bios* (2012, 8).

Secondly, Burgat attempts to deconstruct, if I may say so, the opposition between "to live" (as something that renders the apparent simplicity of the animal life) and "to exist" (as something that only belongs to the humans and only deserves the name of existence). Relying on works by Kurt Goldstein, Gilbert Simondon, Maurice Merleau-Ponty, and Georges Canguilhem, she reconsiders the category of life (as *a bios* that one can describe, i.e. write its *bio*graphy; as *a human bios*) in relation to "the structure of the living being—*Leib* (subject of consciousness, organism, individual)—and the way it behaves" (2012, 18). Thus, her "philosophy of the animal condition" (25) not only challenges philosophical commonplaces that petrify animal existence into a pure and *de-individualized* immanency but also shows how the ontological degradation of animals eradicates them from what Simondon refers to as "the animal kingdom" (1995, 157). What is refused to the animals locked up in "farms, laboratories, zoos, and circuses" is "life of relation" and the "openness of any possibility" (Burgat 2012, 26) as a biological necessity of living beings. Only the recognition of this biological necessity gives us access to the psychical life of animals. Burgat takes up the Husserlian distinction between *Leib* and *Körper*. If a *Körper* is a purely physical body and its description, a *Leib*—as a living body, an "animate organism" (Husserl 1960, 97)—is a body inhabited by a psychical ego which is conscious of itself because it lives through this body and can experience the world when touching and being touched. Drawing on the concept of incarnate subjectivity described by Jan Patočka, Burgat says that "the experience of the self is an experience of the world at the same time" (2012, 17). *An* existence, be it animal or human, is simply what belongs to *a Leib*. From this perspective, eating the animal *Leib* may be considered a culturally sanctioned form of primitive cannibalism.

At the final stage of her "philosophical investigations," Burgat analyses the institutionalized human carnivorousness from the perspective of philosophical anthropology. Broadly discussing "practices and representations in which the animal is good to think as long as it is good to eat," she claims, however, that "the carnivorous option, as an institution, is of alimentary and culinary character in a secondary way only" (2017, 356). The institutionalized and culturally legitimized habit of eating meat, as Burgat claims, stems from what Derrida coined the "sacrificial structure," in which the place "for a noncriminal putting to death" is left open (Derrida 1991, 112). Killing (sacrificing) animals should be then reported to this structure which is consubstantial with the subject, who becomes one by "the executions of ingestion, incorporation, or introjection of the corpse," (112) and *driven* by reason. Quoted by Burgat, Horkheimer and Adorno say:

> In war and peace, arena and slaughterhouse, from the slow death of the elephant overpowered by primitive human hordes with the aid of the first planning to the perfected exploitation of the animal world today, the unreasoning creature has always suffered at the hands of reason.
> (Horkheimer and Adorno 2002, 204; Burgat 2017, 32 and 383)

The first traces of rationalized reason, which tends to instrumentalize itself by reifying the unreasoning living being, would have its origin in the advent of humanity with its subjectivity as *hypokeimenon*. As Burgat points out:

> [H]e who fires at everything which is alive is neither the modern man nor the bourgeois. Nor is it the industrialised West only. It is, much more broadly, this man characterised by the subjectivity as *hypokeimenon* and by what is to be found underneath. One may see in this subjectivity the advent of humanity, which considers itself an entity separated from the rest of the living, including cultures whose ontologies are continuist but whose practices are discontinuist.
> (2017, 383)

The edification of the animal industry, were we to believe Burgat, is a consequence of "the primitive overpowering of animals" (Horkheimer and Adorno 2002, 161). This argument, however, needs to be critically discussed.

Establishing a direct link between institutionalized butchery and sacrifice as a "need, desire, authorization, the justification of putting to death, putting to death as denegation of murder" (Derrida 1991, 115), Burgat opposes her argument to the Marxist interpretation of animal exploitation and to what she refers to as a "materialist reading" (2017, 361). According to her, such a reading tends to see in animal exploitation nothing but a craving for profit, while the real issue of institutionalized butchery should be recognized as metaphysical (361). Thus, thinking "with Freud and against Marx" (2017, 361), Burgat argues that the death drive, apart from the carnivorousness of subjectivity and the instrumentality of reason, would be another instance of the sacrificial structure (391), while eating animal flesh (*Leib*) turned into meat would stem from man's primitive inclination to aggression, described by Freud as "an original, self-subsisting instinctual disposition in man, [which] constitutes the greatest impediment to civilization" (Freud 1962, 69). Burgat believes that "abolishing the most patent forms of [the] expression [of this inclination] (private property, profit, and money) is possible, but will not undermine [it], as such an attempt takes on the signifiers and not the signified of violence" (2017, 362). And she asks: "Is eating meat ever seen as a matter for this aggression and a will to annihilate animals?" (362).

Burgat's argument here is strategically understandable. She posits that the sacrificial logic which makes humans carnivorous subjects is impassable metaphysically. Thus, her entire effort is to attempt to consider the equivalence principle within this logic and to dissociate the human metaphysical need to make offerings from the sacrificial structure described by Derrida. Ultimately, despite her anthropological drift towards ancient cultures, Burgat's reasoning is very pragmatic, clear, and down-to-earth: what can be substituted for animal victims in the sacrificial structure is vegetal meat and, more importantly, clean meat (also referred to as cultured meat or in vitro meat). The technical invention of clean meat makes it possible to consider the equivalence principle within the sacrificial structure, which sustains the human craving for meat and saves animals from our anthropologically institutionalized carnivorousness. And yet, when she opposes Marx with Freud by demonstrating the inalienability of the sacrificial structure which overdetermines carnivorous humanity, Burgat entirely relies on Freud's reservations about "the communists [who] believe they have found the path to deliverance from our evils. According to them man is wholly good and is well-disposed to his neighbour; but the institution of the private property has corrupted his nature" (1962, 59–60). However, this communist assumption, Freud continues, "is founded on an untenable illusion. By abolishing private property one deprives the human love of aggression of one of its instruments, a strong one undoubtedly, but assuredly not the strongest" (60).

Let me put this bluntly: Freud's argument is shallow. In fact, we do not know who exactly "the communists" are, here represented in a deliberately caricatured and rather ironic way. The repetition of this argument, by replacing more-or-less important "instruments" with "the signifiers" and "the signified," seems problematic, even though it ultimately leads to a convincing hypothesis whose political potential is undisputable. Despite all my enthusiasm for Burgat's discovery, I argue that, thinking of, against, and—last but not least—from within carnivorous humanity, we should not think "with Freud and against Marx" but rather *both with Freud and Marx*. The reasonable approach to animals from a perspective critiquing the political economy, which I opt for here, does not at all blame the unreasonable capitalist drive to profit from animal exploitation, as Burgat seems to suggest when referring to "some contemporary authors" whose perspective would be limited by their "materialist reading" (2017, 361). No one knows who these authors are, yet they strangely resemble Freud's communists.

In my consideration of animals with (a certain) Marx, which tends to see them as objects of a critique of political economy, I would consider cultured meat a *good* disruptive innovation about to come. I call this innovation disruptive because, once applied to the economies of scale, it would destroy the meat industry and create a new market for meat-eaters. If such a disruption would not change our dietary habits in a substantial way, the social change this disruption is supposed to provoke would be *radical* in the deepest meaning of the term. In fact, the change, according to the capitalist logic of every disruptive innovation, would have to affect the very roots [*radix*] of what we have referred to as society since the birth of sociology, which took place within the already industrialized Western world. The politico-economic possibility of rebuilding and transforming ties of friendship between humans and animals within the *polis* has not been lost, provided that we oppose what Stiegler describes as "an economy of carelessness [*économie de l'incurie*] on a global scale, founded on a systemic extension of stupidity [*bêtise*]" (2015, 3–4), and we have *in mind* that the forgotten face of this carelessness is not only the systematic exploitation of fools (*les bêtes*) but also that of animals (*les bêtes*). To put it as simply as possible, we need to stop being economic fools if we want to struggle for a human-animal future. The reinvention of the human-animal bonds of *philia* (Lauxerois 2002, 89–100) would be an attempt to re-articulate the "zoopolis" (Donaldson and Kymlicka 2013) within the *digitalopolis*. Thus, I would be eager to approach cultured meat as a positive disruptive innovation that might deliver animals from their captivity in hyper-industrial society,[4] closely connected to the ongoing debates about the end of work due to automatization, universal basic income, or, what Stiegler refers to as, the economy of contribution.

My consideration of animals with (a certain) Freud, on the other hand, tries to retrieve the *primitive* meaning of Freud's *Wunsch* (both desire and wish) (Lichtenstein 2017, 182) to even oppose it to Freud's *Todestriebe* (the death drive) that Burgat refers to in her analysis of the sacrificial structure. Drawing upon Stiegler's attention to *technē*, strictly related to the atrophy of desire in the consumerist capitalism, my approach to animals puts an emphasis on the fundamental role of techni-*city* in animal exploitation. In fact, those primitive human hordes who overpower an elephant have already fabricated *instruments* for doing so. Their life is technical, while their techni-city is what distinguishes them from other living beings.[5] From this perspective, the origins of animal murder would be purely technical: it would stem from the desire to transform in/animate matter rather than from destructive and aggressive instincts.

I am not saying that my hypothesis totally excludes that of the sacrificial structure, although the Freudian notion of *Todestriebe* as one of the fundamental instances of this structure seems to me quite problematic. I rather attempt to shed light on what Burgat seems to have underestimated in her analysis of the carnivorous fact. This technical perspective makes it possible to describe a radical shift which took place in the transition from technical instruments to technological machines, on the one hand, and in the transition from the industrializing world to hyper-industrial society, on the other. Any attempt to describe the animal condition, I argue, should not neglect these transitions. In fact, if animal murder has its origin in human technical inventiveness, which made it possible to reify animals and ab/use them as animate matter subject to be *eternally* transformed and de-formed, only technology at the service of an unreasonable economy could de-animalize them as unreasoning beings. A *technical* distinction between the animal reification and de-animalization is necessary to understand why the invisibility of processed animals by means of the general intellect is unprecedented in "the history of the notion of animal life" (Simondon 2004, 29) and why it should be referred to as what Stiegler describes as the systematic and systemic stupidity of the "there is no alternative" ideology (2013, 2015).

4. Towards Animal Dreams and Desires

There is every reason to believe that animal murder was intrinsic to the foundation of the *polis* (city), since the killing of unreasoning beings was *originally* possible by the techni-city of human, reasoning life. *However*, the shift from the reification of animals to their de-animalization by means of the general intellect may be seen as a poisonous effect of *technē*, described by Stiegler in its pharmacological structure, on the life of the spirit (*esprit*), which is tantamount to the life of the mind (*esprit*). In fact, when overpowered by *technē*, which made de-animalization possible, man, having lost the capacity to see (*idein*) the animal form [*idea*] of life, is the one who suffers from a deficiency of the soul in relation to what Aristotle described as the *psyche* in his *Peri psyche* (*On the Soul*), where "he included psychology in biology" (Simondon 2004, 52). Introducing *Deux leçons sur l'animal et l'homme* de Simondon, Jean-Yves Chateau reminds us that *psyche*, referred to as *anima* in Latin, is

> what animates, the life principle, were it human, animal, or vegetable. One is alive when one moves by oneself, when the principle of change and that of movement (or their absence) is in oneself by essence and not by accident (as opposed to technics).
>
> (2004, 6)

When she observes "the *non-naturality* of the animal life" (2012, 15), Burgat, unlike Agamben (2004, 13–14), clearly posits that the main distinction to be made is between the life of plants and that of animals (humans included), and not between the life of animals and that of humans. Having this *in mind*, one must say that noetic life—which Stiegler attempts to heal in his practice of "positive pharmacology" (2015, 33)—cannot come into existence from the principle of *noesis*, that is to say, from animal life.

If there is a metaphysical issue in institutionalized butchery, this issue should not be linked, as Burgat claims, to Freud's *Todestriebe*. I would rather link this issue to *stupidity as mindlessness*, claiming that it is necessary to reconsider this mindlessness in the context of the technological development of capitalism and a critique of the political economy. This necessity to struggle against mindlessness by means of the part of the *psyche* which distinguishes the human from the animal assumes that *Toderstriebe* is the best example of the neoliberal economy destroying psychic life (Stiegler 2015). *Toderstriebe* does not involve killing unreasoning beings; it rather systematically and systemically destroys what Freud defined as *Wunsch*. Crudely speaking, the general intellect driven by *Toderstriebe* tries to make us *unthink*, that is to say, to make us *dis-animated*, by making us believe that a desire for change is impossible, while the destiny of the hyper-industrially processed animals is sealed. Yet we are not allowed to content ourselves with our own desires or dreams. It is necessary to refer the Freudian notion of *Wunsch* to Aristotle's biologico-psychic approach to *psyche* in order to attempt to rethink *Wunsch* as an irreducible threat of every self-aware living being and, as a result, to reinvent the Aristotelian category of *bios*. Perhaps, having made us pay attention to the animal *Leib*, Florence Burgat, all in all, has blazed a trail for understanding animal desires and dreams.

Notes

1. This chapter is part of a research project "In the Name of Friendship, for the Sake of Community: Friendship and Community in Contemporary French Thought," financed by the National Science Center (NCN), Poland. No. 2015/17/D/HS2/00512 (Sonata).

2. If not stated otherwise in the reference list, all translations from French are my own.
3. "Disruptive innovation is a term in the field of business administration which refers to an innovation that creates a new market and value network and eventually disrupts an existing market and value network" (quoted after Wikipedia). And yet, "disruptive" comes from the Latin verb *dirumpere* (or *disrumpere*): "to dash to pieces; to break, burst asunder" (www.perseus.tufts.edu/hopper/). In his last book, Bernard Stiegler shows how this wicked logic of disruptive innovation makes us fall in the disruption and *go mad* (2016).
4. Although, as Paul Ariès, Frédéric Denhez, and Jocelyne Porcher point out in their widely commented article published in *Libération*, there is a real danger that the idea of cultured meat might be taken over by big tech companies such as GAFA, while the undisputed fashion for veganism might provoke the use of practices typical of predatory animal industry, such as using hormones for making animal proteins grow faster and antibiotics for avoiding contaminations. Indeed veganism at all cost is as bad idea as a ridiculous claim that "veganism is dangerous" (www.liberation.fr/debats/2018/03/18/pourquoi-les-vegans-ont-tout-faux_1637109). One of the biggest challenges in hyper-industrial society is jointing veganism with environmental care. With eight milliard people to feed progressing automatization and the disruptive logic of capitalism, it is, however, highly unlikely to imagine a better human-animal world out of technological innovation.
5. Although this distinction is arguable to a certain extent. The ability to fabricate instruments is not *specifically* human. Ethology provides us countless examples here.

References

Agamben, Giorgio (2004). *The Open: Man and Animal*. Translated by Kevin Attell. Stanford, CA: Stanford University Press.

Aristotle (2000). *Nicomachean Ethics*. Translated by Roger Crisp. Cambridge, UK: Cambridge University Press.

Baratay, Éric (2012). *Le point de vue animal. Une autre version de l'histoire*. Paris: Seuil.

Baratay, Éric (2017). *Biographies animales. Des vies retrouvées*. Paris: Seuil.

Berger, John (2008). *Ways of Seeing*. London: Penguin Classics.

Berger, John (2009). *Why Look at Animals*. London: Penguin Classics.

Boutang, Yann Moulier (2011). *Cognitive Capitalism*. Translated by Ed Emery. Cambridge, UK/Malden, MA: Polity Press.

Bower, Joseph L., and Christensen, Clayton M. (1995). *Disruptive Technologies: Catching the Wave*. Harvard Business Review. https://hbr.org/1995/01/disruptive-technologies-catching-the-wave.

Burgat, Florence (1997). *Animal, mon prochain*. Paris: Odile Jacob.

Burgat, Florence (2012). *Une autre existence. La condition animale*. Paris: Albin Michel.

Burgat, Florence (2017). *L'Humanité carnivore*. Paris: Seuil.

Cassin, Barbara, Apter, Emily, Lezra, Jacques, and Wood, Michael (eds.) (2014). *Dictionary of Untranslables: A Philosophical Lexicon (Translation/Transnation)*. Princeton: Princeton University Press.

Chateau, Jean-Yves (2004). "Présentation." In *Deux leçons sur l'animal et l'homme*, edited by Gilbert Simondon. Paris: Ellipses.

Derrida, Jacques (1982). *The Margins of Philosophy*. Translated by Alan Bass. Chicago: The Harvester Press.

Derrida, Jacques (1991). "'Eating Well,' or the Calculation of the Subject: An Interview With Jacques Derrida." In *Who Comes After the Subject*, edited by Eduardo Cadava, Peter Connor, and Jean-Luc Nancy. New York/London: Routledge.

Derrida, Jacques (2008). *The Animal Therefore I Am*. Translated by David Wills. New York: Fordham University Press.

Derrida, Jacques (2009). *The Sovereign and the Beast*. Vol. 1. Translated by Geoffrey Bennington. Chicago: University of Chicago Press.

Derrida, Jacques, and Roudinesco, Elisabeth (2004). *For What Tomorrow . . .: A Dialogue*. Translated by Jeff Fort. Stanford, CA: Stanford University Press.

Despret, Vincianne (2012). *Que diraient les animaux, si . . . on leur posait les bonnes questions?* Paris: La Découverte.

Despret, Vincianne, and Porcher, Jocelyne (2007). *Être bête*. Arles: Actes Sud.

Donaldson, Sue, and Kymlicka, Will (2013). *Zoopolis: A Political Theory of Animal Rights*. Oxford: Oxford University Press.

Fontenay de, Elisabeth (1998). *Silence des bêtes. La philosophie à l'épreuve de l'animalité*. Paris: Seuil.

Freud, Sigmund (1962). *Civilization and Its Discontents*. Translated by James Strachey. New York: Norton & Company.

Harfeld, Jes Lynning, Cornou, Cécile, Kornum, Anna, et al. (2016). "Seeing the Animal: On the Ethical Implications of De-Animalization in Intensive Animal Production Systems." *Journal of Agricultural and Environmental Ethics* 29, no. 3: 407–23, https://doi.org/10.1007/s10806-016-9611-1.

Hegel, Georg William Friedrich (2013). *Philosophy of Nature*. Vol. 3. Translated by Michael John. London: George Allen & Unwin Ltd.

Horkheimer, Max, and Adorno, Theodor W. (2002). *Dialectic of Enlightenment: Philosophical Fragments*. Translated by Edmund Jephcott. Stanford, CA: California University Press.

Husserl, Edmund (1960). *Cartesian Meditations: Introduction to Phenomenology*. Translated by Dorion Cairns. The Hague/Boston/London: Martinus Nijhoff Publishers.

Lauxerois, Jean (2002). "À titre d'amitié." In *Amicalité. Éthique à Nicomaque–Livres VIII et IX by Aristotle*, translated by Jean Lauxerois. Paris: éditions À propos.

Lichtenstein, David (2017). "The Dialectic of Desire: A View of Intrapsychic Conflict in the Work of Jacques Lacan." In *Psychoanalytic Perspectives on Conflict*, edited by Christopher Christian, Morris N. Eagle, and David L. Wolitzky. London/New York: Routledge.

Lyotard, Jean-François (1984). *The Postmodern Condition: A Report on Knowledge*. Translated by Geoff Bennington and Brian Massumi. Manchester: Manchester University Press.

Malabou, Catherine (2010). *Plasticity at the Dusk of Writing: Dialectic, Destruction, Deconstruction*. Translated by Carolyn Shread. New York: Columbia University Press.

Marx, Karl (1887). *Capital: A Critique of Political Economy*. Vol. 1. Edited by Frederick Engels. Translated by Samuel Moore and Edward Aveling. Marx/Engels Internet Archive. www.marxists.org/archive/marx/works/1867-c1/.

Marx, Karl (1973). *Grundrisse: Foundations of the Critique of Political Economy (Rough Draft)*. Translated by Martin Nicolaus. London: Penguin Books in association with New Left Review. www.marxists.org/archive/marx/works/1857/grundrisse/.

Rancière, Jacques (2004). *The Politics of Aesthetics: The Distribution of the Sensible*. Translated by Gabriel Rockhill. London: Continuum.

Simondon, Gilbert (1995). *L'Individu et sa genèse physico-biologique*. Grenoble: Millon.

Simondon, Gilbert (2004). *Deux leçons sur l'animal et l'homme*. Paris: Ellipses.

Stiegler, Bernard (2003). *Aimer, s'aimer, nous aimer. Du 11 septembre au 21 avril*. Paris: Galilée.

Stiegler, Bernard (2005). *Constituer l'Europe 2. Le motif européen*. Paris: Galilée.

Stiegler, Bernard (2010). *For a New Critique of Political Economy*. Translated by Daniel Ross. Cambridge, UK/Malden, MA: Polity Press.

Stiegler, Bernard (2011). *The Decadence of the Industrial Democracy: Disbelief and Discredit*. Vol. 1. Translated by Daniel Ross and Suzanne Arnold. Cambridge, UK/Malden, MA: Polity Press.

Stiegler, Bernard (2013). *What Makes Life Worth Living: On Pharmacology*. Translated by Daniel Ross. Cambridge, UK/Malden, MA: Polity Press.

Stiegler, Bernard (2014a). "LPharmacologie de l'épistémè numérique." In *Digital Humanities. Organologie des savoirs et technologies de la connaissance*, edited by Bernard Stiegler. Paris: FYP éditions.

Stiegler, Bernard (2014b). *The Re-Enchantment of the World: The Value of Spirit Against Industrial Populism*. Translated by Trevor Arthur. London: Bloomsbury Academic.

Stiegler, Bernard (2014c). *Symbolic Misery, Volume 1: The Hyper-Industrial Epoch*. Translated by Barnaby Norman. Cambridge, UK/Malden, MA: Polity Press.

Stiegler, Bernard (2015). *States of Shock: Stupidity and Knowledge in the 21st Century*. Translated by Daniel Ross. Cambridge, UK/Malden, MA: Polity Press.

Stiegler, Bernard (2016). *Dans la disruption. Comment ne pas devenir fou?* Paris: Les Liens Qui Libèrent.

Stiegler, Bernard (2018). *The Neganthropocene*. Translated by Daniel Ross. London: Open Humanity Press.

Wismann, Heinz (2012). "Avant-Propos." In *Une autre existence. La condition animale*, edited by Florence Burgat. Paris: Albin Michel.

5 Laika's Lullabies

Post-anthropocentric Representations of the First Dog in Space[1]

Anita Jarzyna

1

This chapter focuses on what, at first sight, may appear to be a marginal chapter of the complicated history of space flights, namely, the use of non-human species in space exploration. In particular, it will analyse the representations of the animal experience in official and popular culture, as well as in historiographic and artistic narratives, choosing, as a case study, the story of Laika—a female dog, and the best known of all the animal passengers sent in so-called "missions" during the Space Race between the Soviet Union and the United States. None of the versions of this historical event can be told without mentioning a few facts and dates that privilege the human perspective.[2]

To celebrate the 40th anniversary of the October Revolution, on 3rd November, 1957, the Soviet Union launched a human-made satellite, Sputnik 2, to carry the first living being outside of the Earth's atmosphere. This living being was a three-year-old mongrel female dog called Laika. After capturing her on the streets of Moscow, scientists trained Laika in the laboratories of the Institute of Aviation Medicine. Nikita Khrushchev ordered the execution of this risky endeavour to take place only a month after the launch of the first human-made satellite in history, Sputnik 1 (on the 4th of October). He kept the project secret, and even out of the schedule of the Soviet Space Programme, in an attempt to give the USSR an advantage over America in a race that was part of the Cold War. The Russians prepared the flight in a great hurry—just a few weeks—fully aware of the faulty engineering that such a tempo could entail. The main issue was that there were no technical solutions to enable the safe return of the satellite's passenger. The organizers planned from the onset that Sputnik 2 would not return. Laika died a few hours after lift-off, contrary to the plan—according to which she was to die after about a week—but she was the first subject of telemetric transfer data on the reaction of a mammal to cosmic radiation, or to the lack of gravitational force. The scientists assumed that the information would be more reliable than that provided by animal tests conducted in laboratories and that it would soon enable the launch of humans into space.

The way in which Laika's fate is presented and commemorated in most of the artistic and historiographical accounts of this event is eloquent enough; as either a hero or, more rarely, a victim of space flights, Laika is almost always inscribed into narratives that presuppose an anthroponormative image of the community. That is, we include animals in our community and treat them as individual subjects only exceptionally, when they play an important role in human history. The search for an alternative perspective must involve a new framing of the co-presence of species, one that exceeds the discourse of "merit" and is expressed in a new language. This new language is represented by the emancipatory and critical language of poetry, which challenges the dominant narrative of animal space flights and distances itself from the anthropocentric perspective that objectifies non-human protagonists. Two renowned Polish poets of the same generation, Zbigniew Herbert (1924–1998) and Adam Włodek (1922–1986), both authored critical poems right after Laika's space flight,[3] resisting the hierarchical norms of human-animal relations and calling for an appreciation of the dog's subjectivity. The poems testify to the fact that both poets were deeply moved by the event, even though it did not resonate so strongly in Poland as it did in the USSR.

To fully illustrate the significance of Herbert's and Włodek's precursory gestures, I first intend to briefly outline the historical accounts of the "soviet space dog programme" over the last 60 years. Polish literature offers nearly no narratives of the programme, while English-language humanities research offers extensive multidirectional studies on the subject, analysing the contemporary propaganda narratives and illustrating their reception in the collective imagination and popular culture. I will only consider Russian-language sources through English-language studies, which extensively quote soviet documents and historiographic materials which are otherwise unavailable. In the last part of the chapter, I compare Herbert's and Włodek's poems to a remarkable graphic novel recently published by Nick Abadzis, *Laika* (2014). This work is neglected in studies dealing with the representation of space animals in popular culture but deserves attention due to the interesting "biography" of the dog that it presents. By artfully linking documents with fiction, it demythologizes the story of this famous dog, as the aforementioned poems by Herbert and Włodek do.

2

The year 2007 saw the publication of a monumental monograph entitled *Animals in Space*, a book as informative and indispensable as it is problematic for animal studies. Quite surprisingly, and contradicting the work's title, Colin Burgess and Chris Dubbs[4] declare in their introduction that the central topic of the monograph are the people who used the animals, not the animals themselves:

> In our pursuit of space we have asked animals to involuntarily lead the way and demonstrate safety, risks and hazards that should never be first identified by a human space traveller. This book records the animals' contributions, but, more importantly, it recognises the thousands of scientists, engineers, veterinarians, programme directors and managers, technical staff, administrative personnel, animal care staff, flight support personnel, astronauts and cosmonauts, who have made space an attainable goal in an inhospitable environment.
>
> (Burgess and Dubbs 2007, xxviii)

We cannot overestimate the value of this work, which reconstructs with incredible detail the history and prehistory of space flights using animals (not only mammals but also insects and reptiles). Nevertheless, Burgess and Dubbs remain interested primarily in the advances of technical solutions ("the death of Laika—they affirm—overshadowed the incredible accomplishment of the launch of Sputnik 2" (Burgess and Dubbs 2007, 164)), and their narrative is clearly based on the anthropocentric calculation, according to which we are somehow obliged to sacrifice animals for the sake of humankind. Of course—they add—we should do whatever we can to treat animals with dignity, but ethical questions raised by space programmes remain mostly outside the scope of their investigation, which, for instance, focuses on animal physiological reactions during experiments to the exclusion of their possible emotional responses. On the contrary, Burgess and Dubbs try to prove that animals have never been reduced to objects in laboratory experiments, since people and scientists would consider these animals as "astronauts." That is why the authors make extensive use of terms like "space dogs" and "space monkeys," even creating the neologism "animauts."[5]

We can observe a similar narrative framework in the research conducted by Olesya Turkina, a Russian art historian and author of a popularizing book entitled *Soviet Space Dogs*. This work is especially valuable for its collection of commented iconographic materials testifying to the global renown of the "space dogs" and their place in the popular culture of the Soviet Union and its satellite states, and even on the other side of the Iron Curtain. Turkina, in fact, seems to support and reinforce the mythologizing narrative that also appears in the book of Burgess and Dubbs:

> [S]pace dogs and space monkey were not simply victims sacrificed for the sake of a scientific experiment. They were perceived as real colleagues by the scientists, *loved companions who could only be sacrificed as much as the scientists were willing to sacrifice themselves.*
>
> (Turkina 2015, 207–8)[6]

By defining animals and scientists as colleagues or even companions, the author certainly elevates the status of animals (they are not simply victims but rather co-workers), and yet she does not give enough attention to the asymmetrical relationship between animals and humans: after all, it was people who used and sacrificed animals in potentially dangerous experiments in order to protect humans from their consequences.

A slightly different perspective comes from the books and articles by Amy Nelson, an American historian who has devoted three generous studies to the Soviet space programme's use of dogs. Nelson's works, even though they do not engage in the history of animals as practised and promoted by Baratay, "address the possibilities of integrating animals into the history of the human past" (Nelson 2011, 134) and also take into account and accentuate Laika's identity and perspective:

> Laika is an extraordinary animal with an extraordinary history. Just as the history of the domestic dog is inextricably bound up with the history of humanity, Laika's story, and that of the space dogs more generally is intertwined with human histories—stories of Cold War rivalry, of exploration and adventure, of loss and sacrifice, and of relationships and dreams. She is famous as a dog, as a pioneer, as a victim, and as a symbol of human aspiration. The anthropomorphic and anthropocentric inflections of her celebrity may say more about the relationship between people and dogs than about the reality of their experience, but they do not undermine the most essential ingredient of her fame and her place in history—her life as a dog.
>
> (Nelson 2010, 220)

First of all, Nelson meticulously reconstructs "the role of dogs as historical actors in a drama that intertwined humans and animals" (Nelson 2010, 205). Focusing on animals allows for a multilateral description of the history of the Space Race as a critical stage of the Cold War:

> Conflicting human perspectives on the first space dog drew on and intensified more generalized tensions generated by the intertwined nature of domestic dog and human ecologies. They also tapped the excitement and apprehension occasioned by the advent of the nuclear era and the Space Age, which suggested the compelling attractions as well as the tremendously destructive potential of technological and scientific advances.
>
> (Nelson 2008, 239)

Secondly, by resorting to sociological concept of the "boundary object," Nelson highlights both the various rhetorics used by USSR propaganda (animals are alternatively defined as "brave scouts," "faithful servants," "experimental animals," "hapless victims," or "heroes") and how "these sometimes divergent meanings converged," since they all contribute to the denial of animals' exploitation.

Thirdly, Nelson's studies permit a better understanding of the impact of Laika's death and the way it was perceived or instrumentalized in public debate. Americans especially—already at work on their space programme with monkeys—along with other Western European countries, used this event against the USSR. The Soviet programme, implemented in 1951, remained secret until 1957, when foreign journalists first learnt about flights with dogs. Already from that time, this news began to inspire isolated protests from the British Society of Happy Dogs (Burgess and Dubbs 2007, 145) and promoters of animal rights, and then even broader dissent. However, the use of animals in space programmes does not seem to have ever been condemned on either side of the Iron Curtain (Nelson 2011, 150). Rather, it was Laika's particular fate that influenced the mass imagination and inspired so much empathy, since she belonged to a species privileged by people and yet consciously sent to death. In this respect, it is worth noting that even critical voices showed a certain degree of ambivalence, insofar as they shared a fascination with the fact that a mammal could successfully be sent into space (Burgess and Dubbs 2007, 163). We should recall that only a few years later, a mission with two dogs, Belka and Strelka, ended with their successful return from orbit. This was received rather positively by the public, even though Belka and Strelka had also been subjected to harsh training and experimentation by scientists (Turkina 2015, 151). The Soviets assumed that Laika's death would have been subsumed into the "ethos of utility" (Nelson 2011, 149)—promoted since Pavlov's experiments on animals, and typical of USSR propaganda—praising the necessity of citizens' sacrifice for the common good. There is no place for household animals in socialist society; a dog should serve the broader community just like a person (Nelson 2006, 123–45, 2016, 43–59). Hence, as Turkina had already noted, "Soviet mass media found itself unprepared for the worldwide concern over the dog's fate, and the outrage resulting from her death" (Turkina 2015, 98).

Laika's death inspired unexpected agitation which was noticeable among USSR citizens. This considerably influenced the way in which other dogs used in space programmes were presented to citizens:

> Rather than deny that the lives or treatment of the dogs mattered (because they were not human), or regarding them as interchangeable, anonymous research subjects, Soviet attitudes towards the space dogs reflected a recognition of canine-human interdependence, and the unique capacities of dogs as "friends" of humanity and "servants" of science.
>
> (Nelson 2010, 213)

The image of dogs presented to public opinion was ambivalent, though unambiguously positive:

> On the one hand, the space dogs' handlers portrayed their charges as "normal dogs," emphasizing their interest in treats, petting, and other "normal dog" behavior. They were described as "quick-witted, obedient, and healthy," suggesting the desired traits any dog lover would seek in a pet. . . . On the other hand, "the most famous dogs on earth," wore flight suits to their press conferences and were photographed standing atop scientific equipment. Cartoons of the dogs suggested that spaceflight was challenging but fun. When journalists viewed television images of the dogs lying inert and helpless in the first moments of weightlessness during the flight, they were told that the dogs were "resting" before settling down to their "breakfast".
>
> (Nelson 2011, 147)

Nelson aptly identifies the implications, not necessarily evident to its recipients, of the double rhetoric of "companions" and "heroes" who sacrifice themselves for science, simultaneously bringing to light the real living conditions of dogs in laboratories, to wit, their traumatic experience. The trusting relationship between dogs and their trainers served to lead dogs through painful experiments, the effects of which were disguised when presenting the ongoing research to the public:

> This tension between regarding dogs as helpmates, servants, and friends and the compulsion to "sacrifice" them for science had complex and often contradictory implications. On the one hand, the dogs were treated and represented as unique individuals. . . . They were given names and interacted with in ways that enabled the people who worked with them to describe their "personality". . . .
>
> On the other hand, . . . contemporary practice turned an assortment of unrelated stray dogs into an array of collaboratively manufactured biotechnologies designed to generate correct scientific data. Dogs selected for orbital flights underwent surgery to have the carotid artery rerouted to the outside of the neck inside a fold of skin. Once the wound had healed, the dogs were trained to tolerate the attachment of a blood pressure cuff to the fold. They also had monitors implanted to enable researches to assess pulse, respiration, and heart function before, during, and after flights. The "space suits" they wore for press conferences concealed the monitoring wires and scars associated with these implants. Underpinning discourses about the individuality of these exceptional but "ordinary" dogs, then, was a powerful impetus to use them as a "technology".
>
> (Nelson 2011, 151–2)

This paves the way for further research which aims to reconstruct the biographies of space animals and tell the history of space explorations without overlooking the identity of all protagonists, human and non-human. To do so, however, it is necessary to collect and confront different kinds of sources, including ethological studies, which would help re-establish, imagine, and verbalize animals' complex experiences and perspectives: how long did it take for these animals to get used to the spacesuits and to the different diets imposed to them? How much pain and suffering did these animals experience?

Lastly, hoping to contextualize Nick Abadzis' graphic novel, I will conclude the chapter with an analysis of another facet of the media representation of Laika which requires emphasis: the depiction of Laika, Belka, and Strelka as superheroes, and their inscription into a fairy-tale narrative. These dogs unexpectedly fulfilled the Soviet Air Force's motto from 1936: the promise to "make fairy tales come true" (Turkina 2015, 10). Soviet propaganda made Laika and other space dogs "simultaneously real and fantastical beings" (Turkina 2015, 13), so that "[t]o Soviet children, the story of Laika was a heroic fairy tale about a kind and intelligent dog that had flown away into Space. To adults, her fate ostensibly resembled their own" (Turkina 2015, 137–8). It is not surprising, therefore, that space dogs quickly appeared in children's books, such as *The Adventures of Belka and Strelka* by Yuri Galperin (1961), *Tyapa, Borka and the Rocket* by Marta Baronova and Yevgeny Veltisov (1962), *Star Travellers* by Yuri Yakovlev (1962), and *Belka and Strelka, and Their Journey* by Maria Poznanskaya (1964).[7] In all these examples, the dogs are attributed self-consciousness regarding the importance of their role in space missions. Interestingly enough, these presentations differed very little from those addressed to adults: animals are given human traits, such as bravery, and this anthropomorphizing is mostly used to dismiss, dissimulate, and justify their abuse. As already stated, the official popular narratives foreground that Laika was the first living being in the Earth's orbit, relegating the fact that she never returned to the background. Likewise, a famous newspaper illustration depicts a dog walking out of a space shuttle and showing people a briefcase filled with results from her space research (Nelson 2010, 212). These animals' cartoon-like depictions appeared on postcards and stamps, matchboxes and packs of cigarettes, sweets and toys—significant memory carriers produced both in the USSR, its satellite states (like Poland), and the West.

Even though mostly focused on the role that animals played in human history and culture, the research I have briefly presented certainly opens a new strain of studies that allow the critical comparison of different sources: photographs prepared for the documentation of the experiments and training, iconographic materials distributed in the press, narratives that circulated during and after the launch of the satellite, and literary and artistic representations of space animals.

My study places itself in the theoretical frame built by these scholars and attempts to give value to three unconventional representations of Laika's fate that put into the foreground the dog's subjectivity. The texts I analyse in the following pages have been neglected by previous studies on the subject, even by the insightful works written by Amy Nelson. When it comes to the artistic reception of the use of animals in space, even Nelson mostly prefers representations in which Laika appears as a sort of pretext, privileging works such as Lasse Hallstörm's film *My Life as Dog* and Jeanette Winterson's novel *Weight*, where the dog primarily becomes a universal figure of loneliness.

3

Nelson argues that in the 1980s alone, "echoes of Laika's immediate celebrity have inspired an array of creative endeavours" (2011, 154). The scholar has gathered an impressive body of texts referring to Laika, especially songs in several languages (2008, 253), but has overlooked the precursory poem by Zbigniew Herbert.

Print on 1st December 1957, in the *Głos Młodych* (The Voice of the Young), "First the Dog" ("Naprzód pies") appeared during the media coverage of Laika's flight, only a month after the Soviet Union sent Laika into Near Earth orbit and 18 days after the official announcement of her death in space, making Herbert one of the very first who popularized her story. Even though Herbert's poetic oeuvre does not reveal in general any particular attention to animals, he immediately reacted to this event with "First the Dog," criticizing the propagandistic narratives of Laika's flight and, *avant la lettre*, also its many future anthropocentric re-appropriations of her story in diverse languages.

Ten years after the first edition of Herbert's poem, Czesław Miłosz and Peter Dale Scott translated the poem and included it in the 1968 English-language selection of Herbert's poetry with the dedication "To Laika" (Herbert 1968, 101)—which was not envisaged in the original edition—probably because they assumed that the American readership would otherwise miss the reference to the dog. This translation was reprinted 40 years later in the English edition of Herbert's collected poems.[8] The new editor, Alissa Valles, decided to leave the dedication, explaining in a footnote who Laika was and in which circumstances she died, though neglecting to note its insertion by the English translators and not the author.

Most of the interpreters of this poem read it as a metaphor of human experience and struggle, disregarding the circumstances in which it was published and treating the story of Laika only as a pretext.[9] Moreover, it should be noted that the introduction of such an unambiguous dedication risks undermining Herbert's intention, which may have been concerned with a broader issue, pointing to the similar abuse of other anonymous animals.

So first the faithful dog will go
and after it a pig or ass
through the black grass will beat a track
along it will the first man steal
who with iron hand will smother
on his glass brow a drop of fear

so first the dog honest mongrel
which has never abandoned us
dreaming of earthly lamps and bones
will fall asleep in its whirling kennel
its warm blood boiling drying away

but we behind the dog *and [behind]* second
dog which guides us on a leash
we with the astronauts' white cane
awkwardly we bump into stars
we see nothing we hear nothing
we beat with our fists on the dark ether
on all the wavelengths is a *whining*

everything *we [one]* can carry on board
through the cinders of dark worlds
name of man scent of apple
acorn of sound quarter of colour

should all be saved for *our [one's]* return
so *we [one]* can find the route in an instant
when the blind dog *leading us [leads]*
and barks at the earth as at the moon[10]
(Herbert 2010)[11]

Herbert's poem stems from the intuition that the implicit meaning of most narratives about Laika's role in the mission of Sputnik 2 was questionable and controversial. "First the Dog" interweaves facts, speculations, and fiction. Let us first consider the sources of Herbert's poetic imagination. The pig, the ass, and especially the second dog that precede man's presence in space are metonymic for the uncountable animals exploited in the ground and air experiments before attempting manned space flights. Quite possibly, Herbert knew about the announcement for the international journalists at the Moscow press conference on 12th November 1957, where the Russian officials noted that Laika had died a week after take-off from oxygen deprivation. Nelson reports that the officials also announced "that developing a way to return space capsules to Earth safely was now a top priority and indicated that many more dogs would fly to space before the first human was sent" (Nelson 2011, 141).

The media regularly covered Laika's health during the seven days after take-off, so Herbert could not have known that she had died as early as in the first five to six hours after the launch due to stress and overheating in the improperly isolated capsule. This information was made public only in 2002. Nevertheless, Herbert surprisingly describes the probable circumstances of Laika's death very aptly when he affirms that "its warm blood boiling drying away" and imagining her "whining." When dying in the cramped capsule, she could have in fact whined or rushed around. However, the scientists received only her vital parameters; no one heard her because the transmitters carried no voice. The first microphones were only set in Belka's and Strelka's cabins. Before the flight of Sputnik 2, Soviet radio broadcasted a transmission with the she-dog barking, which is why she was given the name Laika ("Barker") despite her "real" name Kudryavka, meaning "Little curry." Radio contact with the capsule was lost after ten days and the satellite itself—first a "kennel," then a coffin—crashed in April 1958. Despite all this, Herbert wanted Laika's imaginary whining to surpass the second stanza, in which she dies, to outlast and spread across his poem, as the animal's final complaint. In my view, Laika's "whining" is one of the most important cues of the piece: Laika's lamentation reverberates across the poem, suggesting that the human plans, hypotheses, and calculations were flawed. The only voice coming from space in this poem, then, is that of an earthly creature, a well-known non-human being. Thus, it appears that the orbital flight brought back only information about the suffering of an animal who, on the contrary, was sent to space to retrieve data about the world above. This means that—in the end—the creature was sacrificed in vain in the name of science, only to confirm the *status quo* of the anthropocentric norms of Earth's civilization.

The title of Hebert's poem refers precisely to this issue, the unscrupulous use of animals as mere tools for broadening and reinforcing human power and knowledge. The title is ambivalent in its purpose: on the one hand, "Naprzód pies" ("First the Dog") sounds like a command to which one should not object. Similar to "sit!" or "down!," it cannot represent a true interspecies communication, but it reveals the complete subordination of dogs to humans and their interests. On the other hand, "Naprzód pies," without an explicit exclamation mark, refers to species domination in an indirect and ironic way, since it can also be translated as "dogs first," which recalls the paternalistic and patriarchal "ladies first." Across the poem, "Naprzód pies" comes a second and a third time, with little variation. In the first line of the poem, the conjunction "so" suggests that the phrase is part of a logical cause-and-effect chain or a prepared schedule of activities (a flight schedule, for example) which require action and execution in a specific order. The conjunction seems to suggest that the poem comes to terms with the aforementioned assumption according to which the sacrifice of animals is required for the achievement of a superior goal: to safely substitute the animals with people one day. In this respect, it is worth noting that the first man to make it into Earth's orbit on 12th April 1961, Jurij Gagarin, remarked, "Am I the first human in space, or the last dog?" (Turkina 2015, 31), because he too was not a pilot, but a mere replacement for "livestock" on board (a term used in official documents; "manned flights" appeared only when the participants' actions were a necessary part of the endeavour). However, the title points also to another and more important meaning that contradicts the previous: "Naprzód pies" ("First the Dog") inverts the order of human priorities and invites us to think first and foremost of the dog's interests and well-being. In other words, Herbert's poem reminds us that an animal, forcefully made into an astronaut, remains first and foremost himself or herself. He or she still dreams of tasty snacks, city walks, lampposts to mark with urine, and barking but may no longer bark at the moon.

However, the second stanza seems to introduce an ambivalence concerning Laika's pain and agony. She falls asleep as if lulled by a lullaby, and she seems to die in silence, as if it had been a peaceful death, but this only serves to lull the human imagination and expectation. This is probably the reason why Laika's whining "reaches us" too late, only in the next stanza, even though there is perhaps another reason: this whining situated at this point of the poem serves to introduce the next dog in line for further experiments. In fact, a second dog after the first appears in the poem almost unnoticed, as if they were tools for men to conduct safe but virtual extra-terrestrial journeys via the mediation of machine sensors and cables, although with no sight nor hearing. Until now, everything went according to plan: men sent animals into space, and one of them died just as expected. The first two verses in future tense reveal this regime imposed upon reality. To the contrary, the contingency of the present tense refers only to humans, who have been unable to fulfil the roles of pioneers and conquerors which they had planned for themselves. When people felt blind in space, they discovered (tardily) that a guide dog is equally blind. They are both thrown into the void, none of them being a discoverer of new worlds, and both missing the familiar home world. The poet suggests the similarity of humans and dogs by enumerating a set of memories that would save us from destruction. That is, a stock of homemade preserves for the road, a sort of reserve of talismans to return home, which easily refer to the cognitive abilities of both; this set of memories is elusive yet resistant to gravitational force, attached to the Earth: "name of man scent of apple/acorn of sound quarter of colour." Maybe the poet packs them for themselves and the dog? In my view, by equating the situation of man and dog, Herbert refrains from arguments justifying the abuse of animals, according to which, on the one hand, the fate and devotion of animals are equal to man's but, on the other hand, their lives matter less.

The last three stanzas describe the disoriented human as gradually gaining insight into, and maybe even assuming, the dog's perspective, which has understood nothing of what was happening. The blind person loses the purpose of the space mission (along with the reason for abusing animals) as if the whole journey was but a makeshift mirage, stolen from a child's imagination or a fairy tale. This image emerges especially towards the end of the poem, which appears to be motivated by the feelings of hope, helplessness, and, above all, anxiety of the return, as if to provide comfort with a familiar and homely image. By going beyond the realistic convention, Herbert reveals the only harsh truth resulting from the discovery of a new world, the truth of the oppressive relationship, the breaking of interspecific bonds, with which humans ensnared the dog to send her on exploratory missions, or hold for consolation in the face of empty outer space. It turns out that the bond between people and dogs is—indeed—unbreakable, but only one-sidedly unconditional. From this viewpoint, the single dog convincingly represents the whole species. Laika's whining may also be a hopeless cry for help that stems from her faith in people. Indeed, human blindness testifies to their lack of imagination, whereas the dog's blindness testifies to her unconditional trust.

Ten years after Herbert's "First the Dog," the poet Adam Włodek published a piece for Laika titled "Epitaph for Laika" ("Epitafium dla Łajki"), in which he underlines human short-sightedness rather than blindness, their inability to foresee the consequences of their actions and inventions. Włodek's poem is as exceptional in Polish poetry as Herbert's piece, although the former's work is much less known. Włodek published the poem in the 1968 volume *Kleksy* (Blots), where it was simply titled "Epitaph," but a note in the 1973 selection of his poetry proves that he wrote the poem already in 1958—like Herbert, very soon after the event.

Obviously
you could not have refused
as the rational man asked.
You practised your reflexes. You went to sniff
how things are rolling
in Space.

May you rest peacefully in the emptiness
doggie.

For humans' benefit the atom has been
rummaged through.
For humans' benefit
even formerly
the dynamite.

May you rest peacefully in the emptiness
doggie
  but be at our side
  but be at our side

(Włodek 1967, 27)[12]

In the beginning, Włodek's poem may seem to recall the command we find in the title of Herbert's "First the Dog." In the poem by Adam Włodek, Laika is similarly submissive but seems more active, probably more active than she could have been. Indeed, Laika practised her reflexes and went to space almost as a matter of volition. We know, however, that in Moscow's laboratories, the decisive factor was a dog's natural disposition: calmer and submissive dogs were considered to be suitable for space flights, which means that they were somehow unconsciously "responsible" for their own fate.[13] Therefore, both poets underlined the devotion of their dog-protagonist, as if to remind us of the reasons why the Soviet scientists had decided to use these animals.

Włodek wrote an epitaph for Laika, even though her name appeared only in the second version of the text, as if to recall an event that had meanwhile lost its recognition. The choice of an epitaph is interesting in itself: to celebrate this dog, the author recovers and pursues an ancient but quite neglected tradition of poems, epigrams, and epigraphic inscriptions dedicated to animals (mostly to dogs and horses) which dates back to Greek and Latin literature. This genre breaks the anthropocentric norms in that it attributes honour to individual animals, most of the time seriously and without any ironic intent (see Garulli 2014). However, if Włodek had simply wished Laika a deserved rest after accomplishing a great task for humanity, his poem would be just another example of commemoration solidifying the anthropocentric order—instead, Włodek questions this order. The verse "May you rest peacefully in the emptiness/doggie," repeated twice, cuts the poem into two different parts. It sounds like an enchantment that divides Laika from Earth. Moreover, this utterance separates the seemingly necessary sacrifice of the animal from an ironic reflection on the pernicious consequences deriving from the myth of progress. When it becomes clear towards the poem's end that these supposedly salutary deeds do not have salvific value but instead call for redemption, the prayer *for* Laika becomes a prayer *to* Laika. Włodek suggests that people who use breakthrough inventions without worrying about others' pain should request Laika's intercessory forgiveness. By doing so, Włodek also reminds us that Soviet propaganda aimed to substitute religion with science (the Space Programme). Thus, Włodek admits that the dog-protagonist of his poem fell victim to irresponsible ambitions. After all, without the inventions of dynamite or nuclear fission, there would be no space flight. Włodek suggests that Laika, resting in peace, thrown into the void, becomes a source of human anxiety.

Thus, there is a register for speaking about Laika that breaks the polite anthropocentric forms of commemoration. Herbert and Włodek understood what historians sometimes tend to neglect even today—that Laika's death is first and foremost the death of a living being. Notwithstanding the superior meaning we wish to attribute to the event, we should suspend the meaning applied to Laika's death. These two poems help us understand that real animals, as iconic historical figures, easily become mere curiosities of collective imagination. On the one hand, we perceive them as individuals but, on the other, they gain the qualities of some fantastic creatures. Włodek and Herbert strive to avoid such abuse by trying to foreground and convey Laika's own perspective, describing it as more important than the breakthrough flight of Sputnik 2.

4

In line with the anti-anthropocentric perspective emerging in Herbert and Włodek's poems is the recent graphic novel by Nick Abadzis, *Laika* (2007), which received very good reviews (Foust 2007; Singer 2008) and numerous prizes.[14] Abadzis' intention was to retell Laika's story, avoiding as much as possible anthropocentrism and anthropomorphism,[15] which is why he, unlike most narratives about the dog, does not focus exclusively on Laika's space travel—as if her life could be reduced to this mission—but also on Laika's first years as a stray on the streets of Moscow until her capture and transfer to the Institute of Aviation Medicine. Moreover, aptly combining fiction and facts, the author provides a multifocal narrative. Four points of view are represented: Sergei Korolev's (the chief designer of the Soviet space programme), Yelena Dubrovsky's (who was Laika's official caretaker and trainer), Oleg Gazenko's (Dubrovsky's superior, one of the leading scientists of the Soviet animal space programme), and Laika's. Abadzis reconstructs the heroine's restricted perspective by voicing her possible emotions through the words of her caretaker—Yelena Dubrovsky—who builds a strong bond with the dog notwithstanding the warnings of her supervisors. In Abadzis' fictionalized account, Yelena even tries to save Laika and quits after her death. Abadzis employs these anthropomorphizing measures very aptly and with thorough knowledge of the extent of a dog's cognitive abilities (attachment to humans, joy, sense of loss, fear). Moreover, he strives to convey Laika's perspective without the words or mediation of other characters, but only through hand-drawn images. Abadzis distances his story about Laika from heroic narratives and refuses to defend the people who decided Laika's fate, but he nevertheless portrays their own attachment and care for the dog, representing their perspectives and justifications. Sergei Korolev is portrayed particularly suggestively as a scientist who, after being imprisoned in a gulag and then fully rehabilitated, sincerely believes in the importance of the mission which also entails the sacrifice of animals.

Laika does not end with the launch of Sputnik 2. We observe how the whole team follows Laika's life parameters, demonstrating to her premature agony. Abadzis not only illustrates the science station but also the inside of the space shuttle and Laika's death from exhaustion. The next few scenes depict the days after the launch, the joy and regret of those who know how the flight ended, along with the reactions of the public to the event. The last two pages of *Laika* show how on 13 April 1958, the shuttle makes its final orbit around Earth and burns to dust in the atmosphere. The graphic novel ends with a black rectangular frame, symbolizing both impenetrable outer space and mourning itself.

Seven years after the publication of the graphic novel, Abadzis published a second edition where he added a new epilogue in which he admits to constantly resisting the temptation of anthropomorphism in the presentation of Laika's story. Moreover, he adds five alternate endings in the annex, having Laika survive in the first four of them. In the first scenario, thanks to a secret plan of her guardians, her capsule is released before Sputnik crosses the Earth's atmosphere, allowing her to land safely and clandestinely. Another rescue comes by accident, as Laika falls ill at the last moment and is replaced by another dog who dies in space. The next two endings embrace even more fantastic plots. In one of them, Laika reaches a faraway planet still unknown, finds traces of another civilization, and acquires new cognitive abilities. In the other, Laika is exposed to cosmic radiation, which gives her supernatural strength, allowing her to return to Earth and take her vengeance. The fifth ending is completely different, comprising nine frames which follow the thoughts of a man (the author's alter ego) sitting at his drawing desk, and the graphic novel ends with his meditations:

> Thing is, why should anybody worry about the death of a dog from more than half a century ago? There are animals, and people, all over the planet treated inhumanely every single day, as they have been since Laika flew./Take your pick of circumstances, wars and political regimes—the world's still full of unseen and unreported misery. What does the life and death of one dog so long ago mean in the overall scheme of things?/In 2008, the Russian authorities unveiled a memorial statue to Laika nearly fifty-one years after her flight. *They were rightly proud of their trailblazing cosmodog*. She is rendered in bronze upon a rocket shaped like a human hand. Flowers are often placed around the statue./She was the first space traveller. Her journey heralded the dawn of the age we live in—of telecommunications, of smart phones, of the "global community." Everything we take for granted in the modern age began with the flights of the Sputnik satellites./She may have been an unwitting pioneer but *her journey was heroic*./The details of why Laika was sent, how she died and what the world was like then shouldn't be forgotten. And neither should the hope for a new and better age that her flight marked, for all inhabitants of this planet./*That's why she matters. She represented and represents us all* . . . just a humble little dog, trained for the mission she flew, yet who didn't understand what was happening to her./ We all stand on the shoulders of those who made sacrifices, or were

> sacrificed whether they agreed to it or not. It's the reason the luckiest of us live freely./Is there an alternative ending to that?
>
> (Abadzis 2014, 224)

The new endings represent a possible answer to the pleas of many readers who could not come to terms with the sad and unjust fate of the dog. This is a feeling that even the author had shared, since the last metanarrative fragment is introduced by Abadzis' reminiscence of his childhood fantasy about Laika saved by aliens: "Maybe she'd live out her days on the planet Fabuluxxus, lifespan massively extended, faithful companion of Lumino, the Dogstar" (Abadzis 2014, 223).[16] Undoubtedly, the four fictional scenarios are aimed to bring comfort. At first sight, the fifth scenario greatly distances itself from the others, because it sticks to reality, reporting the way in which Laika's story was interpreted and justified in the name of human progress and interests. Although it seems to invalidate the other four fantastic plots, it in fact follows their consolatory tone by insisting on the superior meaning that Laika's death assumes for humankind. Laika becomes a representative of all people. In this way, we all partake in her suffering, but are we sure that we are also able to assume responsibility for her pain? "These dogs were the pioneers for human kind" (Turkina 2015, 14), writes Olesya Turkina in a similar manner. This is why the monument commemorating Laika depicts her proudly standing on a human hand. When we remember Laika in such a way, we offer our gratitude, but not remorse, and above all, we do not cease or challenge the practices that killed her.

The same mechanism functions in the narratives (recorded by Burgess, Dubbs, and Turkina, and retold by Abadzis) about the care that the dogs received in the laboratories, which were their first homes off the streets. The truth is that, undoubtedly, most men and women in white gowns (veterinarians, engineers, and other specialists engaged in the Soviet space flight programme with animals) did all they could to comfort the dogs and satisfy all their needs—in breaks from training and experimentation, that is. There can be no doubts about their attachment to Laika and the other dogs; one of the employees of the Institute of Aviation Medicine even took Laika to his home right before the final preparations for the mission of Sputnik 2, so that she could spend some time with his family and children. Before that, another employee adopted Tsygan, a dog used several times for the space flight programme. Yet these noble gestures changed very little but rather confirm the "rule" organizing the animals' lives, according to which humans establish for them the best possible conditions that would not interfere with research. This "rule," however, requires an explanation of the animals' sacrifices. Some years after the launch of Sputnik, Oleg Gazenko, the man responsible for the medical portion of the training and experimentation, expressed regret for Laika's death: "Work with animals is a source of suffering to all of us. We treat them like babies who cannot speak. The more time passes, the more I'm sorry about it. We shouldn't have done it. We did not learn enough from the mission to justify the death of the dog" (quoted in Turkina 2015, 131, and also reported by Abadzis).[17] This is why we cannot speak of Laika as an exception to this rule without casting doubt on the space programme's core assumptions regarding living beings, as if the problem could be reduced to the hasty decision of sending the dog to space when the available technologies did not allow her to come safe to the Earth.

The fact is that space flights have never become a common experience of humans and animals, they did not bring us closer to each other, and they were not met with an effort to overcome the human/inhuman dichotomy. Nelson writes that "the space program . . . brought humans and dogs together in decidedly unequal ways in an effort to overcome not just the 'great divides' of human/ nonhuman and nature/ culture, but also the forces of gravity" (Nelson 2011, 135). However, what remains crucial is the fact that animals and people participated in the space programme under unequal rules, meaning that Nelson's oppositions were overcome only in a purely physiological sense.

This is why the only proper ending of Abadzis' graphic novel is probably the one that radically exceeds anthropo-normative schemes: the ending without comforting alternatives or justifications, presented in the first edition of *Laika*: a large black frame that covers almost the whole page, an ending that suspends all justifications and calculations, as if discrediting the heroic narrative of individual sacrifice—be it humans' or dogs'—for the common good. It is precisely in comparison to this polyphonic epilogue of five alternative scenarios that the original closing gesture gains even greater significance. Nevertheless, it is worth noting that Abadzis did not delete this first ending, but he rather added an appendix, so as to suggest that humans cannot prevent themselves from re-coding animal lives into fictional narratives that justify the current structure of species domination.

Is there an alternative ending to that?—asks Abadzis' alter ego. Probably not, but as the authors analysed in this chapter demonstrate, there are certainly alternative versions of space flights that contest anthropocentrism and put animal lives into the foreground, allowing a different understanding of our common history on this planet.

Notes

1. This chapter is part of the project "Post-koiné. Zwierzęta i poeci (studia wybranych przykładów w literaturze polskiej)," financed by the Narodowe Centrum Nauki, decision number DEC-2014/12/S/HS2/00182. It is a reworked version of a chapter included in my monograph *Post-koiné. Studia o nieantropocentrycznych językach (poetyckich)* (Studies of Non-Anthropocentric Poetic Languages; Łódź 2019).
2. I refer here to the difference between human and animal history of animals as introduced by Éric Baratay (2014).

3. Even though it is possible to identify some common features (e.g. both poets avoid avant-garde experimentation), their differences in terms of poetics, political orientation, and international recognition are quite considerable. For this reason, it is unusual to see these two authors set beside one another. Nevertheless, their poems about Laika and space missions with animals converge on many points. Markedly, these poems hold a marginal position in the oeuvres of both Herbert and Włodek, as they do not seem to pay special attention to the issue of interspecies relations.
4. Earlier in 2003 Dubbs published short children's book about first dogs in space (Dubbs 2003).
5. "During these pioneering years, the dogs of the Soviet space programme and the monkeys and chimpanzees of the U.S. space effort were treated very much like astronauts, undergoing vigorous training, subjected to danger, and garnering lavish media attention, just as their human counterparts would in the years to come. They became objects of national pride and pawns in the ongoing political gamesmanship of the Cold War Space Race" (Burgess and Dubbs 2007, 373).
6. All markings in the chapter by A.J.
7. An insight about the character of this kind of narrative emerges from the quotations used in Turkina's book (Turkina 2015, 28, 72), which shows, among other things, that dogs eagerly and consciously sacrifice themselves, treat space flight as an unusual adventure, and enjoy the experience of zero gravity. Notably, the 2010 full length animated film for children, *Space Dogs* by I. Evlannikova and S. Ushakov, supports this image. Moreover, the official American narrative about sending animals into space during the Space Race assumed a similar position to the Soviet one. For instance, see the 1962 children's book *The Monkey in the Rocket* by J. Bethell and S. Leone (1962).
8. Hence, "First the Dog" functions in the English literary field for more than half a century. For instance, Marjorie Garber refers to the poem in her book *Dog Love* (Garber 1997, 71–2).
9. The reviewers of Herbert's volume *Studium przedmiotu* ("A Study of the Object"; Łapiński 1961; Kwiatkowski 1961; Lipski 1962), where his poem "First the Dog" appears, do not even mention this particular piece. Markedly, Jerzy Kwiatkowski highlights that "Herbert is the poet of compassion," and he adds: "the range of Herbert's pupils is actually very broad: it more-or-less encompasses this incomplete species that scholars call homo sapiens." Kwiatkowski completely ignored at least one representative of a non-human species which Herbert introduces with great sympathy. For more interpretations of the poem "First the Dog"—all of which omit the fate of the real dog—see Kaliszewski (1990), Wojtyła (2008), Żywiołek (2009), and Szewczyk-Haake (2018).
10. I have marked those parts of the translation which require either additional commentary or correction and may influence its interpretation. Firstly, the English verb "whining" is a translation of the Polish "skomlenie," which is hardly translatable as it denotes the whimpers and other sounds produced mainly by canines. Secondly, the translator uses the personal pronouns "we," "our," and "us" in the last stanza, whereas the Polish original markedly offers the impersonal form, because it may mean both the animal as the dog and the animal as the human-astronaut. A minor error appears to me in the translator's decision to omit the repetition of the preposition "behind," which recalls the progression of time. Instead, the translator changes it into a conjunction that may quite well indicate a simultaneity of the events at hand.
11. Więc naprzód pójdzie dobry pies / a potem świnia albo osioł / wśród czarnych traw wydepczą ścieżkę / a po niej przemknie pierwszy człowiek / który żelazną

ręką zdusi / na szklanym czole kroplę strachu // więc naprzód pies poczciwy kundel / który nas nigdy nie opuścił / latarnie ziemskie śniąc i kości / w swej wirującej budzie uśnie / zakipi—wyschnie ciepła krew // a my za psem za drugim psem / który prowadzi nas na smyczy / my z białą laską astronautów / niezgrabnie potrącamy gwiazdy / nic nie widzimy nie słyszymy / bijemy pięścią w ciemny eter / na wszystkich falach jest skomlenie // wszystko co można w podróż wziąć / poprzez ciemnego świata zgorzel / imię człowieka zapach jabłka / orzeszek dźwięku ćwierć koloru // to trzeba wziąć ażeby wrócić / odnaleźć drogę jak najprędzej / kiedy prowadzi ślepy pies / na ziemię szczeka jak na księżyc (Herbert 2008).

12. The poem is translated by the author of the chapter. Oczywiście / nie mogłaś odmówić / skoro człowiek rozumny prosił / przećwiczyłaś odruchy pojechałaś przewąchać / jak się kręcą sprawy / w Kosmosie // niech ci lekką pustka będzie / suczko // na pożytek zbawienny poszperano / w atomie / na pożytek zbawienny / jeszcze dawniej / dynamit // niech ci lekką pustka będzie / suczko / ale wstaw się / ale wstaw się / za nami (Włodek 1967).
13. Éric Baratay writes about nonviolent resistance as the agency of animals, especially dogs, draught horses, and dairy cows. Though we may consider this approach to be an abuse of sorts, due to the many etiological unknowns, I say that it is worth the risk.
14. 2008—won the "Best Publication for Teens" Eisner Award and was nominated for the "Best Reality-Based Work" Eisner Award.
15. In "A Short Interview With Nick Abadzis" (2007), Abadzis declares that his intention was to avoid anthropomorphism in portraying the central canine character.
16. Interestingly, Owen Davey's (2013) children's book *Laika the Astronaut* (Templar Publishing, Dorking) offers a similar happy ending: here the dog is saved and adopted by a family of extraterrestrial beings. On the other hand, the other recent children's book about Laika, Jeni Wittrock's *First Hero in Outer Space: Laika the Space Dog*, not only does not anthropomorphize her but actually faithfully tells her story.
17. For more, see the valuable study by Jane Johnson (Johnson 259–273), who analyses the situation of animals in space programmes through the lens of vulnerability. She proves that the immanent fragility of the condition of animals is not even considered by people. Moreover, Johnson argues that—compared to the field of biomedical research—arguments about the greater good seem here insufficient and unconvincing.

References

Abadzis, Nick (2007). *Laika*. New York: First Second.

Abadzis, Nick (2014). *Laika*. New York: Square Fish.

"A Short Interview With Nick Abadzis" (2007). *The Comics Reporter*, September 1. www.comicsreporter.com/index.php/resources/interviews/9884/.

Baratay, Éric (2014). *Zwierzęcy punkt widzenia. Inn wersja historii*. Gdańsk: Wydawnictwo w Podwórku.

Burgess, Colin, and Dubbs, Chris (2007). *Animals in Space: From Research Rockets to the Space Shuttle*. New York: Springer Praxis Books.

Davey, Owen (2013). *Laika the Astronaut*. Dorking: Templar Publishing.

Dubbs, Chris (2003). *Space Dogs: Pioneers of Space Travel*. Lincoln: iUniverse.

Foust, Jeff (2007). "Review: Laika." *The Space Review*, September 1.

Garber, Marjorie (1997). *Dog Love*. New York: Touchstone.

Garulli, Valentina (2014). "Gli epitafi greci per animali. Fra tradizione epigrafica e letteraria." In *Memoria poetica e poesia della memoria. La versificazione epigrafica dall'antichità all'umanesimo*, 27–64. Venezia: Università Ca' Foscari.
Hallstörm's, Lasse (1985). *My Life as Dog* [film]. Sweden.
Herbert, Zbigniew (1968). *Selected Poems*. Translated by Czesław Miłosz and Peter Dale Scott. New York: The Ecco Press.
Herbert, Zbigniew (2008). *Wiersze zebrane*, oprac. edytorskie Ryszard Krynicki. Kraków: Wydawnictwo a5.
Herbert, Zbigniew (2010). *Collected Poems*. Translated by Alissa Valles, additional transl. by Czesław Miłosz and Peter Dale Scott, introduction by Adam Zagajewski. London: Harper Collins e-books.
Johnson, Jane (2015). "Vulnerable Cargo: The Sacrifice of Animal Astronauts." In *Commercial Space Exploration: Ethics, Policy and Governance*, edited by Jai Gailliott, 259–73. London: Routledge.
Kaliszewski, Andrzej (1990). *Gry Pana Cogito*. Łódź: Wydawnictwo Łódzkie.
Kwiatkowski, Jerzy (1961). "Zbawienie przez prostotę." *Życie Literackie*, nr 46: 4, 9.
Łapiński, Zdzisław (1961). "Próba współczesności." *Tygodnik Powszechny*, nr 45: 4.
Lipski, Jan Józef (1962). "Między historią i arkadią wyobraźni." *Twórczość*, nr 1: 111–14.
Nelson, Amy (2006). "A Hearth for a Dog: The Paradoxes of Soviet Pet Keeping." In *Borders of Socialism: Private Spheres of Soviet Russia*, edited by Lewis H. Siegelbaum, 123–45. Basingstoke: Palgrave Macmillan.
Nelson, Amy (2008). "The Music of Memory and Forgetting: Global Echoes of Sputnik II." In *Remembering the Space Age: Proceedings of the 50th Anniversary Conference*, edited by Steven J. Dick, 237–53. Washington, DC: The NASA History Series.
Nelson, Amy (2010). "The Legacy of Laika." In *Beastly Natures: Animals, Humans and the Study of History*, edited by Dorothee Brantz, 204–24. Charlottesville/London: University of Virginia Press.
Nelson, Amy (2011). "Cold War Celebrities and the Courageous Canine Scout: The Life and Times of Soviet Space Dogs." In *Into the Cosmos: Space Exploration and Soviet Culture*, edited by James T. Andrews and Asif A. Siddiqi, 133–55. Pittsburgh: University of Pittsburgh.
Nelson, Amy (2016). "Bringing the Beast Back In: The Rehabilitation of Pet Keeping in Soviet Russia." In *Companion Animals in Everyday Life: Situating Human-Animal Engagement Within Cultures*, edited by Michał P. Pręgowski, 43–59. Basingstoke: Palgrave Macmillan.
Singer Jason (2008). "Laika by Nick Abadzis." *New York Press*, October 14.
Szewczyk-Haake, Katarzyna (2018). "Portret rodzinny albo Jakub i Ezaw (*Naprzód pies*)." In *"Nie powinien przysyłać Syna." Etyczne i metafizyczne aspekty twórczości Zbigniewa Herberta*, edited by Józef Maria Ruszar, 95–108. Kraków: JMR Trans-Atlantyk, Instytut Myśli Józefa Tischnera.
Turkina, Olesya (2015). *Soviet Space Dogs*. Translated by I. Cannon and L. Wasserman. London: Fuel.
Winterson, Jeanette (2009). *Weight*. Edinburgh/New York: Canongate.
Wittrock, Jeni (2015). *First Hero in Outer Space: Laika the Space Dog*. Minnesota: Picture Window Books.

Włodek, Adam (1967). *Kleksy*. Warszawa: Czytelnik.

Wojtyła, Konrad (2008). "Odcyfrowana katastrofa albo psia wędrówka w zaświaty." *Pogranicza*, nr 3: 13–18.

Żywiołek, Artur (2009). "Pan Cogito w galerii cyników." In *We mnie jest płomień. Glosy do Herberta (w 10. rocznicę śmierci Poety)*, edited by Marek Biernacki, 55–65. Bielsko-Biała: Wydawnictwo Akademii Techniczno-Humanistycznej.

6 The Blind Spot of the Plot

Thinking Beyond Human With Karel Čapek[1]

Chiara Mengozzi

Despite the ongoing discussion about the beginning of the Anthropocene, the accuracy of its stratigraphic evidence, and the identity of the *Anthropos* at stake in this neologism, there is broad consensus in the scientific community about the anthropogenic nature of current climate change.[2] In other words, we have seemingly entered a new geological epoch characterized by humans as principal agents of environmental transformation. Since the Industrial Revolution, with a significant acceleration after World War II, the changes imposed by humans upon other living beings, and more broadly upon the environment, apparently witnessed a qualitative and quantitative leap which disrupted the conditions of life on this planet, producing *unpredictable* and *unintentional* effects, the magnitude and direction of which are still to be measured. The anthropogenic factors of climate change (such as the use of fossil fuels, intensive animal farming, and deforestation) are undoubtedly intrinsic to industrialization and the development of capitalism on a global scale. However, the current climate crisis cannot only be understood through this theoretical framework: if the history of capitalism and industrialization pertains indeed to the history of humankind, global warming implies longer-term planetary processes, both geological and biological/evolutionary, which are far beyond human control, albeit determining us a species.

In other words, the temporality of human history "has now collided with the timescales of two other histories, both deep time, the time of the evolution of life on the planet, and geological time" (Chakrabarty 2015, 179; cf. also Chakrabarty 2014).

What lessons can be drawn from the observation of this ongoing collision?

1) It follows from the above that we cannot afford to destabilize the conditions that work as limiting parameters allowing and regulating life on the planet, including our own—like the carbon cycle or the temperature on this planet. Of course, as paleoclimatologists put it, life on this planet will probably recover, but it simply will do it after us, and according to a non-human temporal scale.

2) Ironically, the pursuit of mastery over nature, precisely at the moment when we conceive of it as unlimited, only makes human powerlessness explicit in the face of the massive, unwanted effects of our actions (desertification, flooding, degradation of biodiversity, mass extinctions).

3) Increasing awareness of such powerlessness also leads to the recognition that planet Earth and all its different elements do not represent an inert background that can be transformed at will without consequences, since humans' irresponsible act of transformation is likely to affect and upset the general frame that determines and regulates their very existence. Climate crises therefore urge us to acknowledge the uncanny alterity and proximity of non-human agents:

> The fact that this planet has life and processes that support life (the story of *zoe*)—as Dipesh Chakrabarty reminds us in his *Tanner Lectures* (2015, 183)—only forces on us the recognition that however we strategize, the planet remains a coactor in the processes that will delay or hasten climatic shifts.

The collision between human, geological, and biological histories—he contends—leaves us "with our sense of having been decentred from the narratives that we ourselves tell of this place" (183).

4) This new sense of being decentred turns out to be fundamentally beneficial, insofar as it proves the inadequacy of a sheerly anthropocentric perspective. It also impels us to take seriously into account the existence of other agents with which we need to share and constitute a common world. The climactic and environmental crisis—as Amitav Ghosh argues in his fascinating *The Great Derangement: Climate Change and the Unthinkable* (2016)—is not merely a cultural crisis, posing a great challenge to our desires and lifestyles, but also a crisis of imagination and the imaginary, particularly that inherited from modernity. Overcoming anthropocentrism is far from an easy path inasmuch as it is part of our culture at least since *Genesis*' injunction to master all living beings. However, the idea of a planet as something to be tamed, entirely measurable, controllable through statistics, and submitted to probability calculations (which is also an idea interiorized, in terms of narrative structures, through the regularity of bourgeois life represented in modern novels) is a rather recent invention.

This tension between the necessity to think beyond the human and its basic difficulty—which is partly innate and partly historically and geographically determined—will emerge in the following interpretation of *The War With the Newts*, written and published by Karel Čapek in 1936, when the Czech writer, journalist, and playwright had already earned a significant international reputation (cf. Ort 2013). On the one hand, I will try to show how the content and formal aspects of his text may be re-organized around the problematic relation between human and non-human animals; on the other, I will interpret in a symptomatic manner the resistance shown by all the interpreters of the novel to the anti-anthropocentric experience at the core of *The War With the Newts*, insofar as they all end up reading Čapek's novel exclusively as an allegory of the oppression of man by man, e.g. of the rise of Nazism.

A brief summary of *The War With the Newts*, focused on the particular human-animal relations portrayed in the novel, is first necessary. In a remote corner of the Indian Ocean, not far from the Indonesian coasts, Van Toch, the captain of the "famous" Czech merchant navy, discovers by chance a species of medium-sized blackish amphibians, able to walk upright on two legs, and intelligent enough to interact with men. A tacit agreement between these animals and the captain is quickly reached: while the former will procure pearl oysters, Van Toch will provide them with knives so they can defend themselves against the sharks. The news of the discovery soon reaches Europe. The Newts or Salamanders (since humans will henceforth identify these animals as both)[3] become an object of curiosity in European zoos and circuses and are scrutinized by prominent scientists who try to establish their origin and classify them in the animal kingdom, vivisected in laboratories, subjected to atrocious experiments aimed at measuring their intelligence and physical resistance, and transported and traded as a cheap workforce. In short, the Newts unwillingly become an integral part and an engine of the capitalist economy, to the point that the European nations will compete for their ownership in a bitter struggle for the control of resources and means of production. Meanwhile, the number of newts grows exponentially. Their rapid growth, however, is not merely quantitative. After acquiring the rudiments of language and creating a political and social organization, the Newts begin destroying coasts and continents in order to extend the surface of seas and oceans, thereby threatening the very survival of the human species.

In broad outlines, this is the narrative content of *The War With the Newts*. This renowned modernist novel belongs to the sci-fi/apocalyptic vein of Karel Čapek's production, which began in 1920 with the play entitled *R.U.R. Rossum's Universal Robots*, about the revolt undertaken by organic artificial beings called Robots, which immediately propelled Čapek onto the international scene.[4]

The relative simplicity of the story, moving from the Newts' enslavement to their status as the main threat of humanity, contrasts with the extreme complexity of the narrative form. The novel not only simultaneously takes place in several places of the world but mobilizes a gallery of characters so vast that the very possibility of classifying them as main and secondary is compromised. It also constantly alternates narrative voices and focalizations (eyewitnesses, tourists, zoologists, palaeontologists, historians, journalists) and makes use of an impressive array of genres, registers, and styles: parodies are made of colonial fictions, stereotypical scripts of the emergent Hollywood film industry, political pamphlets, academic conferences, newspapers, governmental dispatches, minutes of businesses, and scientific reports.

Of the novel in question, there are many interpretations, whose similarities are relevant to pinpoint here: a preliminary summary of their shared assumptions will lead to the focus-shifting exercise at stake in this chapter.

First, all the interpreters read the novel through an allegorical lens. The Newts would be nothing but a pretext to show the contradictions and the destructive forces of contemporary civilization; "like a mirror, they reflect the image of human values and the contemporary state of culture" (Klíma 1985, xix; cf. Bugge and Hemelíková 1999).

Second, with regard to the contents of the allegory, the analyses suggest three main referents. In the first part of the novel, the salamanders would represent all the "wretched of the Earth": colonized people ravaged by Europe, an underclass mercilessly exploited by capitalism, Africans traded in slave ships, lynched Afro-Americans, nation-state-deprived Jews, in other words, all the minorities who face inequality and discrimination. In the second part, they would embody the mediocrity and the utilitarianism of the "Mass Man," while in the very last part of the novel (the one that has drawn most readers' attention), the salamanders would have been used by the author to satirize the rise of Nazism, since they pursue the expansion of their *Lebensraum* (living space) as a consequence of their excessive proliferation (Van Crugten 1995).[5]

Third, as regards the form, interpreters have unanimously stressed the "polyvocality and poly-discoursivity" (Richter 2011, 208) of the novel as a strategy enabling Čapek to multiply his critical targets such as militarism, capitalism, nationalism, and colonialism. Nevertheless, they all overlook two major stylistic features whose meanings will emerge shortly: the "surprise effect" in the concatenation of events, and the frequent inversion between the centre and margins of the text. In the first case, what is announced at the end of a chapter as a probable development of the plot is usually proved wrong in the next. Even immediately before the Salamanders' uprising, journalists and experts confirm that "world peace is not actually threatened" (Čapek 1985, 169),[6] that thanks to "this peaceful Newt work for the peace and prosperity of all nations . . . never before has the world been closer to perpetual peace and to tranquil though glorious flowering than at this moment" (169). As for the textual inversion between centre and periphery—as testified by the growing disproportion of footnotes, which can occupy several pages in a row—this is also implied by the plot's shift of focus onto the secondary characters and places. Thus, Prague ironically stands as the ideal centre of a narrative that reports the fortunes and misfortunes of global capitalism. On the one hand, Prague is one of the last cities to be flooded by the Newts, and hence it suffers the disastrous after-effects of progress with a certain time lag; on the other, it is precisely in a Prague-based office that the destiny of humankind is decided, as a follow-up to the meeting between captain

Van Toch, the explorer who discovered the Salamanders, and Bondy, the businessman who will initiate the intensive exploitation of these animals on a global scale. Such is also the case with the doorman Povondra—the only character, among many, to be present from the beginning to the end of the novel—who first takes pride in introducing captain Van Toch to Bondy, before wishing he had not. Ironically, the naïve doorman will later view this encounter as the primary cause of the probable extinction of humankind.

Without denying the accuracy of the allegorical interpretations previously mentioned, I would like to show the extent to which they are reductionist and symptomatic of the obstinacy of humankind to consider itself as the centre of the history (here meaning both History and story). I call them symptomatic insofar as they persist in replacing man at the centre of the plot, even when confronted with a radically anti-anthropocentric novel that stages the catastrophic consequences of the anthropocene, i.e. the geological epoch where the ancient human aspiration to become *maître et possesseur de la nature* seems to be accomplished:

> on one side a rising civilization, and on the other eternal and unchanging nature. That barrier, dear ladies and gentlemen, is now falling. . . . What a fantastic experience! . . . only by the emergence of an oceanic culture . . . has our planet become truly and wholly civilized.
>
> (162)

This is what the mayor of Nice states in his introductory statement to a lecture that a particularly evolved and cultivated salamander is asked to give to a congress of eminent scientists. This long digression recalls Kafka's *A Report to an Academy* published in 1917.[7] In this short story, an ape named Red Peter is captured in Western Africa and comes to understand that the only way to elude the miserable fate humans have in store for him in a zoological garden is to become like them. Through his mimicry, he undergoes a metamorphosis so complete that, when asked by the "honoured members of the Academy" to give a talk about his former life as an ape, he is unable to even recollect it.

The fact that most of the analyses of the novel remove the salamanders as referents and re-code them within an anthropocentric symbolic order—although most of the novel, at least until the final overturn, meticulously describes all possible forms of violence inflicted upon these animals (in zoos, circuses, hunting expeditions, and scientific laboratories)—can be in turn interpreted as a clear symptom of the readers' failure and inability to look and think beyond humanity.[8] In other words, before being an allegory of colonized people, the working class, Afro-Americans, Jews, and, in the end, even Nazis, the Salamanders should be considered first and foremost, in narrative terms, as concrete living beings that humans, "through a monstrous symbiosis of barbarity and scientific rationality"

(Magris 2001), classify, subjugate, and treat as if they were insensitive and pain-free, if not merely inanimate objects.

If we resist as much as possible the impulse to allegorize the Salamanders and see them as a mere mirror of intra-human conflicts, the different parts of the novel and its numerous critical targets reveal their intrinsic coherence and complicity. Furthermore, Čapek's novel will prove to be much more contemporary and disturbing than has ever been thought, insofar as it can also be read as a narrative on the unpredictable consequences resulting from the uncontrolled manipulation of non-human living beings, and not (only) as a warning against the rise of Nazism, i.e. a problem hopefully relegated to the past. This was understandably a major concern for Czech readers in the second half of the 1930s, but the time has come to lay aside this canonical interpretation, which is far from representing the only possible meaning of the text, or its overarching message.

"Do not think that the evolution that our life has arrived at is the only possible evolution on our planet. That was it. Every responsibility dates back to this sentence. For this reason I wrote *The War With the Newts*"[9]—Čapek affirms in a short essay which later became the preface of the book. This anti-essentialist, anti-anthropocentric, anti-teleological Darwinian premise is at the origin of the novel: if human beings along with all animal species are the contingent outcome of natural selection and adaptation to changing environments, "it cannot be excluded—as Čapek puts it—that under other favourable circumstances, a different kind of life, let's suppose, another animal species different from humans might have become a vehicle of cultural evolution."[10] In *On the Origin of Species* (1859), Charles Darwin compares artificial with natural selection: humans are certainly less powerful than nature, but through domestication, they are able to impose new conditions upon species causing new inheritable variations. Along these lines, Čapek's novel imagines that the unprecedented encounter between humans and the Newts can be the main driving force for the accelerated evolution of the latter. However, before acquiring the main human attributes and imposing themselves as the new masters of the planet, the Newts pass "through the various institutions in which animals are enclosed, gazed at and tormented" (Richter 2011, 211). In zoos and circuses, they are at first the objects of exotic curiosity, before they are left to die of disease, hunger, or overfeeding; the unregulated, mostly illegal hunting of wild salamanders, is "reminiscent in many respects of seal hunting expeditions in the past" and "tacitly regarded as a natural regulation of the Newts' population" (130):

> Sometimes a dragged body would get caught between the stones; then the sailor would give a violent and vicious jerk and a leg would come off.... On board the captured Newts were flung into tanks.... With the Newts thrown in, it looked thick and repulsive like some kind of noodle soup.... On the following morning four men arrived

> with long poles and poked around in that "soup" (it really is called soup in the trade); they stirred those densely packed bodies and identified those which did not move or whose flesh was falling off; these were then fished out of the tank with boathooks. . . . [E]ach time six to eight items of "damaged goods"—as they called it—were thrown overboard. Our ship was accompanied by a procession of big and well-fed sharks.
>
> (131–32)

In scientific laboratories, distinguished scientists endeavour to measure the Salamanders' psychic life through vivisection, demonstrating their inferiority to humankind. Experiments testing their resistance to cold, dry conditions, chemicals, and mutilation are carried out in order to assess their exploitation as a workforce:

> In the case of one experimental animal, we amputated all its four extremities and its tail: within thirty days it was again complete. . . . [I]t is a matter of some interest that a Newt whose tongue was removed forgot how to talk and had to be taught afresh. If a Newt's head is cut off or its body severed between its neck and pelvic bone, the animal dies. On the other hand, it is possible to remove its stomach, part of its intestines, two-thirds of its liver, and other organs without damage to its vital functions; it can therefore be stated that an all but eviscerated Newt is still capable of life.
>
> (139–40)

Even their meat might be consumed by humans: "when eaten raw it causes acute pain, vomiting, and sensual hallucinations," but "these harmful effects disappear if the cut meat is scalded with hot water"; "it can be boiled or steamed, and will taste like inferior beef. In this way we consumed a Newt we used to call Hans; it was an educated and clever animal" (140).

These excerpts and many others—surprisingly disregarded by most interpreters—exemplify the deep complicity between the scientific discourses produced on the Newts and the practices of subjugation. On the one hand, the various forms of life cannot be disciplined without producing a knowledge that objectifies them; on the other, any knowledge on these living beings, far from producing an objective understanding, strengthens human control over them and proves to serve specific national, political, and economic interests:

> It came about that eventually the natural history of every nation had its own giant salamanders, and was waging a furious scientific war against the giant salamanders of other nations. As a result, that whole

> important business of the salamanders was never sufficiently resolved on the scientific side.
>
> (p. 78).

In Čapek's novel, an astonishing and disparate array of knowledge is deployed, but the proliferation of discourses on the Newts ends up preventing humans from experiencing a genuine encounter with this uncanny alterity. The reader, for instance, will never learn the meaning of the "dance of the Salamanders" (a major rite, which probably preludes mating), since humans attack and slaughter hundreds of them at this very moment when they are at their most vulnerable. The reader will not even know whether it is appropriate to call them Newts or Salamanders.

Indeed, scientific tests turn out to be absurd and tailored to human standards. The theories presented by zoologists and palaeontologists on the animals' mysterious origin are mostly mystifications: the most preposterous hypothesis dwells on the idea that the salamanders' skeleton matches that of an antediluvian man, identified in a fossil discovered in 1726 by the famous Swiss naturalist Johannes Jacob Scheuchzer. Finally, the articles related to the Salamanders, published in newspapers around the world and patiently collected by the doorman Povondra, are either fragmentary (unbeknownst to her husband, Mrs. Povondra from time to time randomly burns some of the documents in the fireplace to endure the harsh Czech winters) or partially incomprehensible (often written in unknown languages). In other words, the discourses produced and spread by journalists, scientists, and historians of different nations provide us with nothing but dubious information about the origin and life of these animals. By contrast, they perfectly inform us on the functioning of the anthropological machine (cf. Agamben 2002), namely the device allowing the recognition of the human through the articulation of a fundamental dichotomy between the human and the animal, simultaneously inside and outside the very category of human. As Jean-François Lyotard put it, "every damage is like a wrong, and turns [the animal] into a victim *ipso facto*," since "the animal is deprived of the possibility of bearing witness, according to the human rules for establishing damages" (1988, 28). Because violence against animals is ethically admissible, the animalization of minorities turns out to be an effective mean of oppression. The allegorical readings of the novel take for granted that the Newts are just a veiled representation of human suffering, but if we insist on the exposure of the anthropological machine within Čapek's novel, then we understand the very reason why the Salamanders are able to represent all sorts of domination of man by man.

In Čapek's novel, the Salamanders not only embody the very paradigm of difference but also delineate the limits of human empathy and blindness towards the cruelties inflicted over other living beings. This is probably the reason why the theme was removed from all the interpretations of the book. Experts, in the name of scientific objectivity, hide the inherent violence of their practices under an asepticized language; the characters involved in the trade of the Newts hypocritically declare that the animals are transported under "humane and hygienic" conditions (128), while millions of them die during the trade after great agony. They also hasten to recall that the sale occurs "in a humane manner and in a way that would not offend human feelings" (103), as if only human interests were at stake—which, incidentally, in Čapek's time, was still the main concern in legislation on animals, aiming to regulate cruelty towards them. To put it briefly, the life of these sentient and intelligent beings is entirely subordinated to their value in use: "just think that every Newt represents some economic value—Bondy the businessman affirms enthusiastically during the general assembly of his enterprise—a workforce value waiting for exploitation" (102).

The only doubts concerning the legitimacy of human behaviour are either raised by secondary characters, such as Mr. Povondra, or relegated to the margins of the text, in the footnotes. At the same time, it should also be noted that Povondra and his family are the only characters who are present throughout the entire plot, and that footnotes, at times several pages long, tend to overthrow the main narrative:

> "I was tormented by doubt—a secondary character affirms in a footnote—on whether we were entitled (in the strict scientific sense) to speak about our (I mean: human) mental life unless we first removed each other's cerebral lobes and cut each other's sensory canals".
>
> (137)

> "Maybe those Newts were only defending themselves", Mrs Povondra asks after the Newts' first attacks. "That's just it" Mr. Povondra muttered uneasily. "Once those brutes start to defend themselves, it'll be a sad day".
>
> (173)

While politicians and experts from all the "civilized" nations of the world assert, through official communications, human mastery over the Salamanders, pretending the planet is not at risk, in Prague—represented as a peripheral capital whose strategic role within international geopolitical interests was, however, perfectly clear to contemporary reader—and, more precisely, in the household of a humble family, Povondra and his wife anticipate further developments of the plot, predicting the imminent fall of the human species.

What if another form of life started treating humans in the same way that humans exploit the Salamanders and other animals? Such is the anti-anthropocentric mental exercise urged upon the reader in the last part of the novel. What if the salamanders, whose extraordinary capacity of adaptation and evolution has been proven, did nothing other than imitate us, reacting to the dreadful conditions imposed by the human species on them? Kafka's ape, Red Peter, declares in his report that "there was no attraction for me in imitating human beings; I imitated them because I needed a way out [*Ausweg*]" (Kafka 1983, 257). Red Peter points to an alternative to captivity in zoological gardens, because he realizes that he had to become like men in order not to end up in a cage.[11] Similarly, the Newts display a mimetic performance as a way out of torture and exploitation: defined as a "caricature of human civilisation" by Mister X—one of the narrative voices of the novel—in a warning addressed to humans, the Newts end up duplicating those aspects of human civilization (speciesism, racism, nationalism, militarism, colonialism) that can assure their survival and enhance their position on the planet at the expense of other creatures.

The hypothesis that another form of life may one day take over is by no means Karel Čapek's invention. The idea is already at the core of several novels written at the end of the ninenteen century, and in particular by one of the founding fathers of science fiction, H.G. Wells, whose works were well known to Čapek, and could be found in his private library. In 1889, H.G. Wells published a novel that met with an immediate success: *The War of the Worlds*. This novel, which would be rewritten and adapted in many ways, gave a narrative canvas to a countless—both literary and cinematographic—production dealing with the invasion of the Earth and the encounter with alien creatures. *The War of the Worlds* casts a gloomy light on (European) man's vanity and assurance of being the undisputed master over life, matter, and the planet:

> And before we judge of them too harshly—as the narrative voice says from the beginning—we must remember what ruthless and utter destruction our own species has wrought, not only upon animals, such as the vanished bison and the dodo, but upon its inferior races. . . . Are we such apostles of mercy as to complain if the Martians warred in the same spirit?
>
> (Wells 2005, 18)

Wells' and Čapek's novels not only develop the analogy between our treatment of animals and that which other living beings might impose on us but also share a general sense of uncertainty about the future of humankind and its supposed mastery over this planet:

> For that moment—as *The War of the Worlds*' narrator puts it when the extinction of humans seems henceforth inevitable—I touched an emotion beyond the common range of men, yet one that the poor brutes we dominate know only too well. . . . I felt . . . a sense of dethronement, a persuasion that I was no longer a master, but an animal among the animals, under the Martian heel. . . . [T]he fear and empire of man had passed away.
>
> (Wells 2005, 124)

> At any rate—the narrator, witnessing the imminent catastrophe, comments at the end of the novel—whether we expect another invasion or not, our views of the human future must be greatly modified by these events. We have learned now that we cannot regard this planet as being fenced-in and a secure abiding place for Man.
> (Wells 2005, 205)

In *The War of the Worlds*, Martians annihilate humans with the same indifference with which humans step on ants. By the same token, in Čapek's novel, the salamanders pay humans back in their own coin. However, the two novels considerably diverge on one important respect. Unlike Wells' novel, where the human treatment of animals does not bear any relationship to human extinction, in Čapek's novel, what threatens the very existence of humans on this planet is not an extrinsic factor (such as the arrival of aliens) but rather the alterations imposed by humans upon the Salamanders and, more globally, upon nature, which they treat as if it were a backdrop against which they could forever deploy their transformative agency.

Such an alternative thematization of Čapek's novel brings focus back onto relations with non-human (animals) regarding other textual details as well, related both to the novel's content and to its form. Two of them deserve special attention. First is the fact that the destruction of coastlines, metropols, and entire continents by Salamanders in search of a marine space proportional to their increasing number manifests itself in the form of natural catastrophes, such as earthquakes, landslides, floods. Second, that one of the narrative voices insists more than once on the idea that no war is actually waged between humans and Salamanders in the strict sense of the word, because war is a human business: Salamanders do not recognize international laws, do not declare wars, nor do they make official statements. Their silence is frightening, their next step unpredictable.

> Let us not flatter ourselves overmuch on account of our human victories over nature—writes Engels in *Dialectics of Nature*—for each such victory it takes its revenge on us. Each of them, it is true, has in the first place the consequences on which we counted, but in the second and third places it has quite different, unforeseen effects, which only too often cancel out the first. . . . Thus, at every step we are reminded that we by no means rule over nature like a conqueror over a foreign people, like someone standing outside nature—but that we, with flesh, blood, and mind, belong to nature, and exist in its midst.
> (Engels 1975, 460–1)

It is precisely this deep and ungraspable interdependency—combined with humans' astonishment, seeing that they are not the only agents on this planet—that manifests itself in the often contradictory chain of events reported by the narrators, in the characters' consternation, blindness, and paralysis when faced with salamanders' agency, in their systematically denied assumptions or interpretations that proceed through trials and errors, in the derisory tone and irony that accompany the likely extinction of the human race. In a more-or-less intentional way, Čapek and his contemporaries' anxiety about the destructive and self-destructive power of Europe, on the eve of World War II, has been stratified in these formal and structural traits of the text. If we follow the narrative logic of the novel and take seriously, or rather literally, its figures of speech, the primary cause of this destructive power is to be found in the immoderate exploitation of the Salamanders, as living beings simultaneously different from and similar to humans, and whose interior life and material existence remain for us fundamentally opaque and impenetrable. The manipulation of the Salamanders—after enduring a long removal and relegation to the rank of a simple figurative pretext under which the author would have disguised all forms of domination and intra-human conflicts—can finally reassert its rightful place within the process of thematic configuration at a time when the urgency of contemporary problems, such as the climatic crisis and the increasing loss of biodiversity, exhort us to a sudden recognition of non-human agents which affect our life beyond our capacity to anticipate and control.

After the salamanders have already sunk one fifth of Europe and are preparing to flood Prague (Prague being the ideal centre of the plot), Karel Čapek enters the novel in the last crucial chapter—entitled "The Author Talks to Himself"—taking the floor to stage a dialogue that proceeds in his divided conscience, and that concerns the eventual end of the history (here meaning both the end of the human world and the end of the plot): "'And you're not sorry for mankind?'—asks his double. 'Couldn't they be made to die out somehow? Maybe by some new disease or through degeneration'" (236–7)—continues his inner voice. However, unlike *The War of the Worlds* by Wells, where Martians, through a *coup de théâtre*, must finally succumb to the viruses of planet Earth, no cheap solution is conceivable in Čapek's modernist novel; no *deus ex machina* will intervene to save humans from the unpredictable consequences stemming from their indiscriminate manipulation of living beings: "too facile, old chap. Why should nature put right what man has messed up?"—protests the author (237). Nevertheless, our spontaneous resistance to conceive of life on the planet *without* and *after* humans is such that dozens of alternative solutions are to be pondered by the author and his double until they reach the hypothesis according to which the Salamanders, who have now become indistinguishable from humans, might engage in a struggle amongst themselves, ending in their own self-destruction, and giving the human race the opportunity to start once again. This, however, is just a speculation. As a matter of fact, the novel remains open-ended:

> "And then?"—the author asks himself
> "I don't know how it goes on."
>
> (241)

In Čapek's famous play *R.U.R. Rossum's Universal Robots*, after the extinction of humans, life will continue with the robots having discovered love and probably acquired the capacity to organically reproduce themselves without the intervention of humans, who, therefore, do not hold the secret of life, as they had pretended to. This outcome, whose idyllic-bucolic tone has been for a long time vilified by most of the interpreters,[12] would instead deserve to be rehabilitated inasmuch it represents a powerful criticism of anthropocentrism. Similarly in *The War With the Newts*, the exercise of decentring the human is taken to its extreme, up to the last threshold of the novel, where the end of humankind (or perhaps of one of its possible historical articulations) does not necessarily correspond to the end of the *(Hi)story*.

Notes

1. This chapter is included in the research project PROGRES Q14 "Crisis of Rationality in Modern Thought" and was supported by the European Regional Development Fund-Project "Creativity and Adaptability as Conditions of the Success of Europe in an Interrelated World" (No. CZ.02.1.01/0.0/0.0/16_019/0000734).
2. Introduced by Crutzen and Stoermer (2000), the notion of the Anthropocene is at the centre of a heated debate. Sceptics aside, the crux of the question is when Anthropocene might have started. Ruddiman (2003) and Kolbert (2010), among others, defend an early Anthropocene hypothesis, when agriculture and domestication began in Eurasia around 11000 BC; others, such as Crutzen and Steffen (2003), Steffen et al. (2004, 2011), dated the beginning of Anthropocene to the invention of steam engine in the late eighteenth century, and see a further acceleration in anthropogenic climate change since 1950. Based on the assumption that only industrialized and high-tech societies are responsible for current climate change, "the use of the species category in the Anthropocene narrative" has also been questioned (Malm and Hornborg 2014).
3. It is so in the English translation of the novel, while in the original Czech version these animals are identified exclusively as "Mloci" (plural nominative of "Mlok").
4. The neologism "robot"—derived from the ancient Czech word *robota* (corvée)—was invented by Karel's brother, Josef. Even though in Čapek's play the term designates biologically artificial organisms, it immediately underwent a semantic shift when the play was staged and adapted abroad, and ended up referring to mechanical artefacts. On this play, cf. at least Nikolskij (1978), Kinyon (1999), and Horáková (2010).
5. Richter (2011) is one of the few essays that tries to go beyond the allegorical frame, by focusing on the novel's Darwinian plot. Regarding the formal features, however, this essay gives priority to the modernist techniques of montage and multiperspectival narrative, without dealing in depth with the decentring effect achieved through the systematic reversal between the centre and the margins of the history. For further in-depth studies of this novel, see Nikolskij (1978) and Ohme (2002).
6. Unless otherwise stated, all quotes from Čapek's novel come from the translation by Ewald Oser (Čapek 1985).
7. There is no evidence to prove that Čapek knew this short story by Kafka, but some textual correspondences invite the reader to consider the hypothesis

that *A Report to an Academy* could have been one of Čapek's sources of inspiration.

8. On the consequences of this human failure to think beyond its own kind on animal lives, and on the possible strategies to overcome it, see Fudge (2002).
9. Translation mine.
10. Translation mine.
11. On the mimetic performance displayed by Kafka's Red Peter, see Norris (1980): "Allegorical readings of Kafka . . . interpret the ape as representing man, as symbolizing man, but not as imitating man. The distinction is crucial because by attributing a textual strategy to the author, to Kafka, but not to the ape, we fall into the anthropocentric trap prepared for us. Kafka's philosophical aim is to devalue and deprivilege reason, an enterprise which places him in the deconstructive tradition of Darwin, Nietzsche, and Freud" (1251).
12. See for example Janoušek (1992).

References

Agamben, Giorgio (2002). *L'aperto. L'uomo e l'animale*. Torino: Bollati Boringhieri.

Bugge, Peter, and Hemelíková, Blanka (1999). "Nemoc a epidemie v dílech Karla Čapka z konce třicátých let." *Česká literatura*, no. 3: 241–58.

Čapek, Karel (1985). *The War With the Newts*. Translated by Ewald Osers. New York: Catbird Press.

Čapek, Karel (2011). *R.U.R. Rossum's Universal Robots: A Collective Drama in Three Acts with a Comedy Prelude*. Translated by David Short. London: Hesperus.

Chakrabarty, Dipesh (2014). "Climate and Capital: On Conjoined Histories." *Critical Inquiry*, no. 1: 1–23.

Chakrabarty, Dipesh (2015). "The Human Condition in the Anthropocene." In *The Tanner Lectures on Human Values*, 138–88. New Haven, CT: Yale University Press.

Crugten, Alan Van (1995). "La Guerre des salamandres: rire et catastrophe." *Études slaves*, no. 2: 5–13.

Crutzen, Paul J., and Steffen, Will (2003). "How Long Have We Been in the Anthropocene Era? An Editorial Comment." *Climatic Change*, no. 61: 251–7.

Crutzen, Paul J., and Stoermer, Eugene (2000). "The Anthropocene, Global Change." *IGBP Newsletter* 41: 17–18.

Darwin, Charles (1890). *The Origin of Species by Means of Natural Selection*. 6th Edition. London: John Murray.

Engels, Frederick (1975). "The Part Played by Labour in the Transition from Ape to Man." In Karl Marx and Frederick Engels, *Collected Works*. Vol. 25, 452–65. New York: International Publishers.

Fudge, Erica (2002). *Animal*. London: Reaktion Books.

Ghosh, Amitav (2016). *The Great Derangement: Climate Change and the Unthinkable*. Chicago/London: University of Chicago Press.

Horáková, Jana (2010). *Robot jako Robot*. Praha: KUP.

Janoušek, Pavel (1992). "Čapkova poetika dramaty a nepovedené konce jeho her." In *Studie a dramatu*, 39–71. Praha: H&H.

Kafka, Franz (1983). "A Report to an Academy." In Franz Kafka, *The Complete Stories*, edited by Nahum N. Glatzer, 250–62. New York: Schocken Books.

Kinyon, Kamila (1999). "The Phenomenology of Robots: Confrontation with Death in Karel Čapek's R.U.R." *Science Fiction Studies*, no. 3: 379–400.
Klíma, Ivan (1985). "Čapek's Modern Apocalypse." In Karel Čapek, *The War With the Newts*, translated by M. Weatherall and R. Weatherall, vi–xxi. Evanston: Northwestern University Press.
Kolbert, Elizabeth (2010). "The Anthropocene Debate: Marking Humanity's Impact." *Yale Environment* 360. http://e360.yale.edu/features/the_anthropocene_debate__marking_humanitys_impact.
Lyotard, Jean-François (1988). *The Differend: Phrases in Dispute*. Translated by Georges Van Den Abbeele. Manchester: Manchester University Press.
Magris, Claudio (2001). "Utopia e Disincanto." In *Utopia e Disincanto: Storie, speranze, illusioni del moderno*, 7–16. Milano: Garzanti.
Malm, Andreas, and Hornborg, Alf (2014). "The Geology of Mankind? A Critique of the Anthropocene Narrative." *The Anthropocene Review*, no. 1: 62–9.
Nikolskij, Sergej (1978). *Fantastika a satira v díle Karla Čapka*. Praha: Československý spisovatel.
Norris, Margot (1980). "Darwin, Nietzsche, Kafka, and the Problem of Mimesis." *MLN: Comparative Literature* 95, no. 5: 1232–53.
Ohme, Andreas (2002). *Karel Čapeks Roman Der Krieg mit den Molchen. Verfahren, Intention, Rezeption*. Frankfurt am Main: Peter Lang.
Ort, Thomas (2013). *Art and Life in Modernist Prague: Karel Čapek and His Generation, 1911–1938*. New York: Palgrave Macmillan.
Richter, Virginia (2011). *Literature after Darwin: Human Beasts in Western Fiction, 1859–1939*. New York: Palgrave Macmillan.
Ruddiman, William F. (2003). "The Anthropogenic Greenhouse Era Began Thousands of Years Ago." *Climatic Change*, no. 61: 261–93.
Steffen, Will, et al. (2004). *Global Change and the Earth System: A Planet Under Pressure*. Berlin/Heidelberg/New York: Springer-Verlag.
Steffen, Will, Grinevald, Jacques, Crutzen, Paul, and McNeill, John (2011). "The Anthropocene: Conceptual and Historical Perspectives." *Philosophical Transactions of the Royal Society A*, no. 1938: 842–67.
Wells, H.G. (2005). "The War of the Worlds." In *The War of the Worlds: Fresh Perspectives on the H. G. Wells Classic*, edited by Glenn Yeffeth, 15–153. Dallas, TX: BenBella Books.

Part II

"I Relate, Therefore I Am"

Human-Animal Encounters

7 Animals and Their Keepers in Belle Époque Postcards of the Jardin des Plantes Menagerie

Maria P. Gindhart

The early twentieth century was a heyday for French postcards, and *cartes postales* from the Jardin des Plantes Menagerie in Paris, the world's oldest public zoo, were especially popular. More than just mementos of a bygone era, these postcards are now extremely useful documents for analyzing the relationship between animals and humans, especially zoo keepers, at this time. In particular, Belle Époque postcards of keepers interacting with animals indicate that Menagerie visitors sought contact with or at least proximity to the animals on display—even if it was vicariously through those who cared for them. Drawing on postcards in the archives of the National Museum of Natural History, of which the Menagerie is a part, as well as those for sale on online auction sites, this chapter considers how this particular genre of visual culture highlighted the bonds between keepers and zoo animals, as well as the hazards inherent in those connections, in early-twentieth-century Paris. While there are no *carte postale* images of zoo animals attacking their keepers, such incidents are referenced in postcard captions, and fleshing out these incidents through other sources gives further insight into keepers, their animal charges, and the rapport between the two.

As Ann C. Colley writes in *Wild Animal Skins in Victorian Britain: Zoos, Collections, Portraits, and Maps*:

> Nineteenth-century descriptions of Victorian British zoological gardens and menageries are remarkable for their multiple accounts depicting the public's eagerness to reach through the bars or wire mesh of cages so as to be able to touch or "caress" the enclosed creatures. . . . Desiring to stroke, pat, embrace, and fondle these exotic animals, children and adults introduced their hands into the dens so as to sense the animal's hide brushing up against their fingers or to feel the lick of the creature's tongue. . . . Drawings and photographs from the period frequently capture these encounters—indeed, it is striking just how often these moments are featured in nineteenth-century pictorial representations of zoos.
>
> (Colley 2014, 121)

Based on evidence from postcards depicting the Jardin des Plantes Menagerie, the same was true in Belle Époque France. In one *carte postale* a man reaches through the bars toward two bison, while in another, a woman extends an arm toward the fencing separating her from the two camels on the other side. Although contact is not actually made, that is clearly the goal of the outstretched arms in both instances.

According to Colley, such acts are "an expression of a larger desire to reach for what is beyond the self" (Colley 2014, 128). For others, they are evidence of a historically situated "zoological imperialism," which was as relevant in the French context as in the British one, whereby zoogoers consciously or unconsciously took part in the conquest of exotic beasts (Ritvo 1996, 50). In terms of the topic of this edited volume, the yearning of visitors to make contact with the animals in the Menagerie—and other zoos—is an example of humans trying to cross, or at least blur, the human-animal line.

While Colley focuses on the concept of touch and the role of zoo visitors, Jean Estebanez has considered the role of distance and proximity at zoos in terms of twenty-first-century *soigneurs*. He discusses the contact that keepers have with animals, which gives them a certain aura in the eyes of the public. Estebanez continues:

> Keepers are those who enter the space that is prohibited to the public. In some ways, keepers are perceived as super-visitors who do not need a substitute for proximity, since they can access the animals. They thus overcome the contradiction inherent in zoos of putting the very object of the visit at a distance.
>
> (Estebanez 2010, 49, 52)[1]

The privileged position held by keepers in the Belle Époque was visualized in numerous Jardin des Plantes postcards showing a *gardien* or *ménager*, as they were then usually called in French, inside an animal enclosure. For example, one *carte postale* of the elephant Rachel shows the pachyderm standing placidly, while the keeper beside her has his hand on her ear. The annotated version in the archives of the National Museum of Natural History dates the card to 1905 and identifies the keeper as Clavel (Ms 3316).

Returning to Colley's list of zoogoers' desires, feeding the zoo animals, as seen in multiple postcards, should be added—with keepers again frequently serving as surrogates in this regard. In many instances, the feeding was made easier due to the ability of the animals to extend appendages outside their cages, rather than the humans having to reach inside. Moreover, as seen in various *cartes postales*, a bear stretching its paw or an elephant its trunk between the bars may have given the Menagerie visitors on the other side the sense that the animals were also eager to have contact. The same is true in the case of a giraffe bending its neck over the grille of its enclosure to receive a treat.

Until the 1970s, the public was widely allowed to feed at least certain animals in zoos throughout the world, and such opportunities, albeit much more limited, continue to occur today. The decision to curtail the practice of allowing zoogoers to feed animals was largely due to the high rates of animal mortality, as animals were frequently fed too much and/or fed the wrong things, as well as to the desire to limit the transmission of diseases between zoo denizens and visitors. However, recent studies have shown that public feedings remain popular with zoo visitors as a means to see animal movement, "get a direct personal response from an animal," and "express a caring and nurturing interest in the animal," and motivations were likely not all that different at the turn of the twentieth century (Kreger and Mench 1995, 147).

Fin-de-siècle postcards of the Menagerie indicate that elephants, for one, received their fair share of treats from visitors, with one *carte postale* showing an elephant extending its trunk through the grille of its enclosure being captioned "The snack of the young elephant."[2] In fact, a 1910 children's book about the Jardin des Plantes offered this advice on feeding the elephants:

> Sometimes the Elephants extend their trunk to one side and the other to ask for food; sometimes also, they raise it and hold it straight, opening their large mouth, waiting for you to throw something in it. Present them rather with a morsel that they can take with their trunk; it is much easier than aiming correctly and reaching their mouth; and although these brave beasts are big gluttons, they take what is given to them very delicately, never by force. They love fruit, and you will surely please them by giving them a banana; they also love sugar, biscuits and bread. . . . The keepers feed them hay, straw, boiled rice, biscuits, bread and green vegetables.
>
> (*Un Dimanche au Jardin des plantes* 1910, 18)[3]

In this regard, it is interesting to note how the young readers were urged to give the elephants delicacies outside their regular diet—in order, perhaps, to make themselves more popular with the pachyderms than the keepers who kept them on a stricter regime.

The situation with the sea lions was rather different, with the same book for children noting:

> The keeper gives them only fish—one whole fish at a time, that they swallow whole. Never try to feed them and never throw bread or anything else in their pond. You would only dirty their water, which they do not like; and it annoys the keepers who must keep this water clean.
>
> (*Un Dimanche au Jardin des plantes* 1910, 46)[4]

Although feeding the animals and cleaning their enclosures are both daily activities necessary for the proper functioning of a zoo (Estebanez 2010,

47), the feeding is perhaps viewed as less of a chore than the drudgery of sweeping, shoveling, and mucking—with the need for extra cleaning provoking the ire of keepers. This is echoed in the subject matter of Jardin des Plantes postcards, as several show keepers feeding the animals while cleaning enclosures is not a highlighted activity.

Multiple postcards show a keeper feeding one or more sea lions. In all cases, the pinnipeds are coming out of the water to receive the fish. And the photographs have generally been taken so that, although fencing is seen in the background, there is no barrier in the foreground, thus providing a sense of being inside the enclosure. Only one known exception includes the fence in the foreground, which, of course, is how the feedings would have been experienced by visitors as opposed to keepers. In two particularly dramatic *cartes postales* offering more close-up views, a sea lion is shown on the faux rocks at one end of the basin rising up to take a fish out of the raised hand of a *gardien*. This showmanship was a form of entertainment that highlighted the ability of sea lions to be trained and suggested a certain level of complicity between human keeper and animal charge. Written accounts further tended to anthropomorphize the animals, although it is unclear whether this humanizing of animals resulted from a sense of affinity or from a human-centric point of view that did not consider the alterity of the zoo animals.

As the popular scientific journal *La Nature*, which erroneously—but not uncommonly—identified the sea lions as seals, explained in 1885:

> One can see daily the friendship that the seals have for their keepers. When they approach, the seals rise out of the water, and give unequivocal signs of friendship and intelligence; and when these keepers distribute their food to them, the seals obey them and execute a series of exercises in the hope of being rewarded with a few extra pieces. They plunge, turn on themselves, climb on their rock and rush into the pool.
>
> (Guyot-Daubès 1885, 322)[5]

In fact, according to *Lectures pour tous* in 1900, the sea lion then in residence in the Menagerie was rightfully called a *chien de mer*—or dog of the sea—because of its attachment to its keeper. As the author noted:

> When it is not frightened by the crowd, it comes out of the water at his call, it arrives limping on the stumps which serve it as paws, takes out of his hand the fish that he holds out, and even, in recompense, kisses him delicately on the cheek.
>
> ("Dans les coulisses du Jardin des plantes" 1900, 979–80)[6]

Unfortunately, there does not appear to be a Jardin des Plantes postcard showing such a reward.

Although two different keepers are depicted in the postcards showing the sea lions being fed, only one, "Ménager Neff," is identified in a caption, and then only on one of the cards (see Image 7.1). This would be François-Xavier Neff, the brigadier of the *ménagers*. The annotated copy of this postcard in the archives of the National Museum of Natural History is dated 1905, and two others in the same collection, dated to 1903 and 1905, identify Neff as the keeper feeding a sea lion. While never named in the postcard captions, the brigadier Pierre Defaux is identified as the sea lion keeper in one annotated card dated to 1907 and in two dated to 1910. The decision to identify Neff in the caption to the one card but not to name him or Defaux in the others is perhaps explained by the annotation on the card—"last portrait of Neff shortly before his death" (Ms 3315).[7] The possibility of the caption being an homage to Neff is all the more likely as he, as will be discussed in the following, was killed while working at the Menagerie by an elephant named Saïd.

The sea lion feedings cum performances were on a fixed schedule, which was published in various guides to the Jardin des Plantes, but they were not the only forms of animal theater at the Menagerie. The elephants, for example, were also trained to a certain degree. One 1904 postcard depicts a keeper, identified as Janet in the annotated archival version, pushing a pachyderm's trunk over his head. The card's caption identifies the pachyderm as Toby, an African elephant (Ms 3315). Coaching an elephant in this manner was designed both to please the public and to be of use to the keepers, as "[t]eaching an elephant to lift its trunk causes

Image 7.1 Postcard, c. 1905.

Source: Collection of the author.

it to automatically open its mouth, which serves not only as an anthropomorphized salute but also provides a means to examine the mouth" (Kreger and Mench 1995, 153). Yet another *carte postale* illustrates a scene analogous to one discussed in an article from 1900 entitled "Behind the Scenes at the Jardin des Plantes." Here, the author wrote of being able to get a more private view of one of the elephants in the morning:

> He then indulges in various amusements with his keeper; on a sign from him, he raises his two front feet on a large tree trunk, and rises to his full height, throwing his trunk back.
>
> ("Dans les coulisses du Jardin des plantes" 1900, 978)[8]

Once trained, meanwhile, the elephants would perform for the public as seen in a variety of postcards and other representations of the Menagerie. For example, one card annotated as dating to 1911 and captioned "The Elephant Kouch preening" shows an Asian elephant with his trunk in the air and his raised front feet resting on a horizontal bar of his enclosure grille (Ms 3316).[9]

The postcard of the keeper training the elephant to place his front feet on the tree trunk dates to 1903 and, as indicated in the annotated archival version, depicts Neff and Saïd. The pair are featured on at least six postcards—one from 1902, four from 1903, and one from 1904 (Ms 3315). All place the viewer inside the enclosure with both the elephant and the keeper, thus giving the viewer a sense of having an insider's perspective. In terms of actual contact and complicity between Neff and his charge, the 1902 postcard (see Image 7.2), which shows the *grand éléphant* touching his

Image 7.2 Postcard, c. 1902.

Source: Collection of the author.

keeper's stomach while a crowd looks on from the other side of the fence, is particularly noteworthy. As with postcards showing elephants reaching outside of their enclosures to be fed by visitors, it is the pachyderm that is seeking contact. And the privileged position of the viewer inside the animal enclosure is underscored by the inclusion in the photograph of zoogoers who are on the outside of the grille looking in. Moreover, there is the additional sense that the pachyderm is perhaps just close enough to swing its trunk to touch the holder of the postcard as it currently does his keeper. While there is nothing ominous about this image, and the contact between the elephant and the keeper appears benevolent, the sheer size differential between the two seems to signal the potential for danger.

Two of the postcards, in which Neff appears to be placing a morsel of food into Saïd's outstretched trunk, identify both the elephant and the keeper in the captions. It is perhaps worth emphasizing that Neff is here characterized as a *cornac* rather than a *gardien* or *ménager*. Thought to derive from Sanscrit or Sinhalese, *cornac* was a word used by the French to refer to elephant drivers and handlers, especially in the context of Asia. The use of *cornac* suggests a certain kind of othering of Neff that can be construed as both positive—*cornacs* were viewed as having an almost mystical ability to control elephants—and negative, as this appellation suggested that Neff was somehow un-French, that he was exotic like his animal charge. It was also an inexact term in this context, considering that Saïd was an African, as opposed to Asian, elephant. That Saïd was an African elephant is made clear in the caption of one of these postcards, which was then reissued (see Image 7.3) after Neff's

Image 7.3 Postcard, c. 1905.

Source: Collection of the author.

death with a modified legend reading in part, "African Elephant Saïd who suffocated his Cornac Neff on 30 August 1905."[10] Especially in the wake of his attack on Neff, the emphasis on Saïd's African origins would have underscored the often more violent natures of African elephants and the fact that they are less easily domesticated than their Asian brethren (Wood 2012, 411, n. 12; 423). At the same time, the reference to Neff's violent fate is belied by the seemingly friendly contact between man and beast shown in the image.

According to press reports, which did vary slightly, Saïd had suddenly become furious and killed Neff by seizing him around the waist with his trunk and either twice hitting him against the grille or throwing him on the ground—perhaps before crushing him against the grille. As discussed in these articles, Saïd had been at the Menagerie for over 22 years when this event occurred, and, throughout most of that time, he had gotten along extremely well with Neff, who was his primary keeper. But the elephant had become more difficult with age and seemed particularly agitated in the days before the unfortunate event. There was even mention that Neff had been forbidden to enter the Rotunda, the building in which Saïd and the other pachyderms were housed ("Un éléphant qui tue son gardien" 1905; "Un éléphant assassin" 1905, 1).

Little is known about Neff, as is the case with most of the *gardiens* from the Menagerie and other zoos. Regarding the Jardin des Plantes keepers in the Belle Époque, the main source of general information is the 1891 report on the Menagerie for the Minister of Public Instruction written by Alphonse Milne-Edwards, who became the director of the Museum of Natural History that year. Milne-Edwards noted that it was difficult to recruit keepers, who worked from six in the morning until five or six in the evening including on Sundays. There were not enough of them, and they were paid less than they deserved. In terms of job duties, they were expected to prepare food, feed the animals, clean their enclosures, let visitors into the Menagerie buildings, and then keep an eye on these zoogoers (Milne-Edwards 1891, 15–16).

Beyond the fact that Neff had been awarded a bonus of 100 francs in 1882 for "the most intelligent care" that he provided to the pachyderms at the Menagerie (Raveret-Wattel 1882, CIX), the few existing facts about this keeper come from the reports of his tragic death in the popular press. He had been employed at the Jardin des Plantes for 25 years and commissioned for 20. A devoted and trusted worker, he had been a brigadier at the Menagerie for eight years at the time of his passing. Born in 1856, he lived in a small apartment at 90, rue de Picpus. He left a wife and three children, of which only the oldest was old enough to work ("Tué par un éléphant" 1905; "Un drame au Jardin des plantes" 1905, 6).

While several other "murderous" elephants were put to death—most famously Topsy, who was electrocuted at Coney Island in 1903—Saïd was spared. Moreover, at least one article tried to justify the elephant's actions. Writing in *Le Naturaliste*, Doctor L. Laloy commented:

> The elephant is a very intelligent and very gentle animal; he is vindictive; but in the present case it is certain that he did not have to take revenge on his *cornac* who loved him and cared for him with devotion. How, then, can we explain this sudden fit of fury? I think it can be attributed simply to captivity. Let us imagine what must happen in the mind of an intelligent animal who for many long years has been enclosed in a restricted space. . . . An attack of nervousness is well excusable.

Laloy went on to discuss "the deplorable hygienic state" in which the elephants found themselves at the Jardin des Plantes and urged that they be placed in more ample conditions resembling their natural habitats as much as possible. He further remarked that the Menagerie was "a traditional institution that serves for the moment only the relaxation of the population" (Laloy 1905, 211–12).[11] Laloy thus raised one of the essential conundrums of zoos—as has long been acknowledged, they are ultimately less about the well-being of animals and more about human entertainment and consumption of the exotic. At the same time, Laloy's rationalization of Saïd's behavior paralleled contemporary theories about the ways in which environment and life circumstances influenced human criminals (Wood 2012, 427–8).

In general, both before and after Saïd's attack, there was interest in explaining the elephant's "rather nasty character." The caption of a photograph in an August 1900 article noted, for example, "One must not blame him: his teeth are not in good shape and nothing embitters like toothaches. Every last bit of Saïb's [*sic*] tusk had to be removed and his *gardien*, improvised dentist, is obliged to monitor his jaw closely" ("Dans les coulisses du Jardin des plantes" 1900, 977).[12] The accompanying photograph shows the colossal elephant and his "dentist," who is none other than Neff. The overall message is that animals, like humans, have individual circumstances and situations that influence their character and behavior.

At the time of Saïd's death in early 1907, *Le Gaulois*, echoing Laloy, noted that captive elephants who killed their keepers "are too unhappy not to be excused these moments of fury." But the newspaper also spoke of this "elephant assassin" in the terms of criminality and penitence that were used in similar instances of bad behavior, both elephantine and human, at the turn of the century (Wood 2012, 407–8). It was claimed by some that "Saïd succumbed to the stinging remorse caused by his crime" against the keeper who had been so kind and attentive to him. The elephant was said to have been overcome by chagrin and to have lost weight in his repentance. Rather than resulting from a trial in a court of law, however, his awareness of the atrocity of his crime had come because "he was allowed to meditate in peace, in the cellular regime to which he had been subjected for a long time." Ironically, the same imprisonment used to explain his violent actions was also credited with creating the conditions in which he could reflect on his crime. In the end, moreover, he was perceived as having been pardoned by the children who visited the Menagerie and greeted him with smiles and treats rather than fear and horror (Tout-Paris 1907).[13]

In terms of Neff's death, one Museum employee emphasized "that he is in no way the cause" ("Tué par un éléphant" 1905).[14] Officially, however, the unfortunate Neff was deemed responsible. An inquest begun by the police commissioner of the *quartier*, M. Bouteiller, found that Neff should have placed Saïd in a security chamber before entering to clean his enclosure. Yet Neff had regularly swept the paddock with Saïd inside, and, as previously discussed, numerous postcards show him interacting and posing with the pachyderm. In addition, the press noted that two other *gardiens* had recently been mortally wounded by Kako the hippopotamus in similar accidents ("Éléphant meurtrier" 1905; "Les méfaits d'un éléphant" 1905, 3). The three deaths in relatively quick succession are evidence of the fact that being a keeper has always been an inherently dangerous job with regular mortal accidents even today (Estebanez 2010, 53). For example, it has been noted that at least nine circus or zoo caretakers were killed by elephants in the 1990s (Wood 2012, 411), and a zoo intern was killed by a lion at the Conservators Center in North Carolina in December 2018, as this chapter was being finalized.

One postcard (see Image 7.4), annotated 1903 in the Museum's archival version, features two hippopotami—"Kako, called the Terrible" and

Image 7.4 Postcard, c. 1903.

Source: Collection of the author.

"Liza, his companion"—with the caption giving the last names of Kako's two keeper victims and the dates when they were attacked.[15] Laudy was assaulted on 16 September 1901, and Lancelle on 1 July 1903. Unnamed in the caption of the postcard is the keeper, at left, sticking his right arm through the bars and appearing as if he is about to toss something into Kako's wide open mouth. In what could, with hindsight, be viewed as foreshadowing of his own death, the *gardien* is none other than Neff, who had taken over from his unfortunate compatriots in caring for Kako and Liza. In this and other later postcards, it should be noted, there is always a barrier between the keeper and the hippopotami, either the grille around the exterior enclosure or the gate in the doorway to the Rotunda interior, which can be seen in the right background of this card. While the postcards in all instances continue to suggest that the viewer is inside the hippopotamus park, in the wake of Lancelle's death, the Menagerie employees were forbidden from entering the space while the animals were present. Moreover, the Museum administration installed a system of sliding doors to protect the keepers from Kako ("L'hippopotame meurtrier" 1903, 1; R. 1904, 4).

Another *carte postale* (see Image 7.5), which exists in both black and white and colorized versions, identifies the hippopotamus as Tako in the caption and notes that he devoured his *gardien* on 17 July 1903. The version in the Museum archives is dated to 1904, and Tako has been amended to read Kako, but the date has not been corrected. The

Image 7.5 Postcard, c. 1904.

Source: Collection of the author.

annotations, meanwhile, further note that the hippopotamus is shown with his new keeper, Jamay, who stands behind the gate between the exterior enclosure and the interior of the Rotunda. The *gardien* is holding what appears to be a rod that he is using either to prop open Kako's mouth or to protect himself from the animal's fearsome jaws. Jamay is also shown behind this gate in a 1905 postcard, while the keeper Bonhomme is in a similar position in another card from that same year. A third card from 1905 shows part of the door and gate, and a keeper's arm, at far right (Ms 3316).

In other postcards, annotated 1903, 1906, 1908, 1909, and 1915 in the Museum's archives, Kako and Liza are shown without a keeper, and none of the captions mentions the *gardiens* that Kako killed. Undoubtedly, while the hippopotami remained star attractions at the Menagerie, the unfortunate *ménagers* were forgotten by the public with the passage of time. Yet Kako's powerful jaws remained a focus of attention, perhaps because of the loss of human life for which they had been responsible. A 1910 postcard depicts the keeper Jeannet peering into the hippo's open mouth. In this case, the hippopotamus from Senegal is standing near the exterior grille of his enclosure, and Jeannet is safely on the other side (Ms 3316). While certain Menagerie keepers learned the hard way that hippopotami were violent and deadly, getting them to open their mouths and bare their sharp teeth appears to have served at least some of the same utilitarian and entertainment purposes as having elephants raise their trunks. Contemporary postcards from other zoos, including Antwerp and Berlin, similarly depict keepers posing with one or more hippopotami with mouths agape. Back in Paris, the popular press was all too happy to report that Kako's jaw measured 60 centimeters and allowed for a dilation of 1 1/2 meters and that his incisors and canines were 18 and 20 centimeters long, respectively ("Broyé par un hippopotame" 1903, 1; R. 1904, 4–5). The distinct emphasis on the powerful jaws of these dangerous animals was undoubtedly meant to give visitors a *frisson*, but the hippopotami, as with the elephant Saïd, were paradoxically deemed as criminals when they actually acted like ferocious animals.

In terms of Kako's victims, less is known about Laudy, who was also referred to in newspapers as Landy, Loudy, Lody, and Lovy. His first name was Émile, and at the time of his death, he was 29 years old and lived at 14, rue Quatrefages. Laudy had just gotten Kako to leave his basin when the hippopotamus rushed at him and knocked him to the ground, or against a partition, with a blow to Laudy's stomach with his muzzle. The *gardien* was in agony for several days before succumbing to his injuries on 22 September 1901 ("Colère d'hippopotame" 1901, 2; "La victime de l'hippopotame" 1901, 2).

Less than two years later, Jean-Baptiste Lancelle, whose last name also appears as Lancel and Lancet, was trying to move Kako from his outside basin to his interior cage. The hippopotamus, whom he had taken care of for seven years, knocked him over and bit him, and his injuries included a fractured pelvis, two crushed thighs, and a perforated bladder ("Broyé par un hippopotame" 1903, 1). One newspaper surmised that Liza was the indirect cause of the *ménager*'s death as Kako was afraid of being separated from his mate. According to this theory, Kako's "brute instinct urged him to remove the obstacle he feared" ("L'hippopotame meurtrier" 1903, 1).[16] Scientific evidence for this explanation is lacking, but Kako and Liza were definitely presented as a pair or couple in multiple postcards, and they had two male offspring while in captivity: Kako II born on 20 January 1906 and Marius born on 15 August 1907. In fact, at the time of Kako II's birth, *La Justice* published a *faire-part*, which the newspaper noted might have been sent had such announcements been used by pachyderms, stating the date and time of the arrival while commenting that mother and child were both doing well ("Echos" 1906). There was thus a certain amount of anthropomorphizing of the hippopotami at the Menagerie, which partially explains the "rational, human-like explanations for their murderous rage, such as jealousy, frustration, or anger" (Wood 2012, 420) when they attacked their keepers.

As with Laudy, details about Lancelle primarily come from press accounts at the time of his death. *Le Petit Journal* noted that Lancelle was an "excellent employee," who, after 31 years of caring for the ferocious animals, had been scheduled to retire the following January. The newspaper also noted that he left behind a widow and four married daughters ("Broyé par un hippopotame" 1903, 1). *Gil Blas*, referencing the "crime of the hippopotamus," reported on Lancelle's funeral, noting that Edmond Perrier, the Museum's director, along with the institution's administrative personnel and employees were all in attendance (Le Curieux 1903).[17] Perrier gave a speech at the graveside of the 61-year-old, and funeral wreaths offered by the Museum and its staff were presented by the *gardiens* ("Obsèques du gardien Lancelle" 1903, 3; "Tué par un hippopotame" 1903, 4). While it is unclear which members of the Museum community would have attended the funerals of other keepers, the deaths and funeral services of *ménagers* who died of natural causes were certainly not reported on by journalists.

Ultimately, it was the hippopotamus rather than either of the keepers whom he killed who warranted the most attention, with accounts of Kako's attacks on his *gardiens* mirroring contemporary popular press coverage of human crimes and murders (Wood 2012, 420–1). As *Le Pêle-mêle*, a weekly humoristic newspaper, noted on 20 March 1904, "Following the example of famous criminals, Kako deserves to have his biography written." While sensationalizing his supposed newfound taste for human blood, the article recounted his arrival in Paris at the age of six months, at which time he drank 20 liters of milk daily, whereas his current quotidian diet was two bunches of alfalfa or straw and 20 liters of flour mixed with 30 liters of water. Part of what made Kako "an important figure," despite his crimes, was the fact that he was worth 25,000 francs, a sum equivalent to the annual salary of 20 Menagerie *gardiens* (R. 1904, 5).[18] Kako's sheer monetary value was undoubtedly one of the reasons why he was allowed to live despite suggestions that such a dangerous animal should be put down ("Tué par un hippopotame" 1903, 4).

A recent French-language children's picture book, entitled *Kako the Terrible* and subtitled *According to a Real News Item That Took Place at the Jardin des Plantes in Paris in 1903*, uses the animal protagonist's real name but calls the keeper Séraphin.[19] This picture book briefly fleshes out the relationship between the *gardien* and hippopotamus, who turns on his faithful servant in the end. While drawing new attention to an actual event by fictionalizing it, the elision of Lancelle—and Laudy—into the fictitious Séraphin serves to erase these actual keepers from memory yet again. At the same time, the book does little to offer a true "animal biography" in keeping with the research of scholars such as Éric Baratay, who have sought "to recover animal lives, to restore their existence, to show their singularity" (Baratay 2017, 31).[20]

Kako the Terrible, written in the third person by Emmanuelle Polack and illustrated by Barroux, begins with Séraphin awaiting a delivery from Africa. When a baby hippopotamus comes out of his wooden shipping crate, he observes the *gardien* with outstretched arms while grunting softly. Séraphin, who names him Kako, is smitten by his new charge, daily bottle feeds him 15 liters of milk (from three goats), and monitors the temperature of his bath so he does not get sick. One image shows the keeper, towel over his arm, about to dip his bare foot into the water of the hippopotamus's basin, much as a father might do for a human child. The focus of the book then shifts primarily to Kako, whose aquatic exploits and giant teeth become a highlight of zoo visits, especially for children. Séraphin is not mentioned for several pages, although he is seen leaning against a grown Kako in one image and bringing him a wheelbarrow full of carrots in another. The next double-page spread shows the *ménager* dancing on Bastille Day and notes that he showed up late for work on 15 July 1903. The following double-page spread states that he was oblivious to the fact that Kako was out of sorts, having been terrified by firecrackers and slept very little the night before. In the final pages, Kako pushes Séraphin to the ground when he enters his enclosure to feed him and then devours the keeper instead of his food! While mentioning the good care that Séraphin had provided the pachyderm for seven years, the text now labels Kako "the terrible" and notes that he was "animated by a brute instinct that finally may never have left him" (Polack and Barroux 2013).[21] One of the goals of this children's book is thus to draw on an actual historical episode to underscore that zoo animals, despite their familiarity with their keepers, are not pets but wild animals.

Overall, turn-of-the-century keepers were generally perceived to be caring and faithful, and justifications were sought when their charges actively turned on them. *Kako the Terrible* suggests that a night of festivities caused the *gardien* to be distracted and the hippopotamus to be irritable, with tragic consequences. Furthermore, in the preface to the book, Michel Saint Jalme, the current director of the Menagerie, stresses that *soigneurs* no longer enter enclosures when dangerous animals are present and that the job has evolved to include developing ways to stimulate zoo animals in addition to feeding them and cleaning their enclosures. He also stresses that contemporary zoo animals are almost all born in captivity rather than captured in the wild (Polack and Barroux 2013). Nevertheless, one could argue that a keeper-eating animal protagonist of a children's book is the stuff of nightmares, capable of causing fear in young readers while never calling on them to question the morality of keeping wild animals in captivity. The fact that the book in no way addresses the effects of confinement on zoo residents is all the more surprising given that this topic was one discussed by Laloy and others in the early twentieth century, at which time *gardiens* were at times referred to as *geôliers* or jailers (R. 1904, 5).

Kako the Terrible draws on vintage *cartes postales* to tell the story. More specifically, it uses both pieces and collages of portions of postcards—including several of those discussed earlier—showing hippopotami at the Jardin des Plantes and other European zoos to form the different images of Kako. A baby when he arrives at the Menagerie, his small form is cut out of a single postcard. As he grows over the course of the narrative, he takes up more space on the double-page spreads, with his body formed by assemblages from a mixture of postcards. Artistically, according to Barroux, "[t]he collages with the old postcards allow a sort of *mise en abîme* since inside my hippopotamus is the picture of another hippopotamus" (Petre 2016).[22] This design approach is also an ingenious way to incorporate actual history into the book, with the postcards capturing a sense of the turn of the century and helping transport readers back to that era. Furthermore, the *cartes postales* help reinforce the moral of the story, for, as Polack has noted, the *gardiens* depicted in them "have no sense of danger, they seem to have forgotten that a wild animal remains a wild animal, even in captivity" (Petre 2016).[23]

The choice of postcards is somewhat misleading, however, as not all are of Kako and/or the Jardin des Plantes. For example, the image of young Kako drinking from a bottle is cut from a postcard showing Marius, the offspring of Kako and Liza, nursing from a goat, and this postcard is used in two other representations of the hippopotamus. Meanwhile, another postcard used multiple times shows a keeper inside an enclosure at the Antwerp zoo with the hippopotami Julie and Brook. Yet another, also from Antwerp and depicting a keeper posing next to a hippopotamus whose mouth is wide open as if he is about to bite him, is utilized three times. While evoking the overall history of hippopotami and their keepers in zoos in the Belle Époque, the somewhat indiscriminate use of them conflates individuals, both animal and human.

While *Kako the Terrible* was both inspired in part by and makes use of period postcards, the *cartes postales* themselves, in addition to serving as souvenirs of a visit to the Jardin des Plantes Menagerie, undoubtedly prompted viewers to imagine what it would be like to be a zoo keeper, enter animal enclosures, and perhaps even touch elephants, hippopotami, or other exotic animals. Largely anonymous, *gardiens* worked long hours for little money. Meanwhile, the zoo denizens often had name recognition and a certain star status but no true sense of individuality and agency. While fuller, albeit still limited, details of the lives of both keepers and their animal charges generally only emerged when the former were tragically killed by the latter, Belle Époque postcards are evidence of the relationship between keepers and the zoo pensioners for whom they cared. Postcards of *ménagers* interacting with zoo animals had the added advantage of providing zoogoers with a surrogate sense of contact with and proximity to animals such as Saïd and Kako without the actual danger that led to the deaths of Laudy, Lancelle, and Neff.

Notes

All translations are those of the author.

1. "Les soigneurs sont ceux qui accèdent à l'espace qui est interdit au public. D'une certaine façon, les soigneurs sont perçus comme des sortes de super-visiteurs qui n'ont besoin d'aucun subsitut à la proximité, puisqu'ils peuvent accéder aux animaux. Ils dépassent ainsi la contradiction qui traverse les zoos en mettant à distance l'objet même de la visite."
2. "Le goûter du jeune Éléphant."
3. "Parfois les Éléphants allongent leur trompe de côté et d'autre pour demander à manger; parfois aussi, ils la relèvent et la tiennent toute droite, en ouvrant leur grande bouche, dans l'attente que vous jetterez quelque chose dedans. Présentez-leur plutôt un morceau qu'ils puissent prendre avec leur trompe; c'est beaucoup plus facile que de viser juste et d'atteindre leur bouche; et bien que ces braves bêtes soient de gros gloutons, ils prennent ce qu'on leur donne très délicatement, jamais de force. Ils aiment beaucoup les fruits, et vous leur ferez sûrement plaisir en leur donnant une banane; ils aiment également le sucre, le biscuit et le pain. . . . Les gardiens les nourissent avec du foin, de la paille, du riz bouilli, des biscuits, du pain et des légumes verts."
4. "Le gardien ne leur donne que du poisson—un poisson entier à la fois, qu'ils engloutissent d'un coup. N'essayez jamais de leur donner à manger et ne jetez jamais de pain ni autre chose dans leur étang. Vous ne feriez que salir leur eau, ce qu'ils n'aiment pas; et cela ennuie les gardiens qui doivent tenir cette eau propre."
5. "on peut voir journellement l'amitié que les phoques ont pour leurs gardiens. Lorsque ceux-ci approchent, les phoques se dressent au-dessus de l'eau, et donnent des signes non équivoques d'amitié et d'intelligence; et lorsque ces gardiens leur distribuent leur nourriture, les phoques leur obéissent et exécutent une série d'exercices dans l'espoir d'en être récompensés par quelques morceaux supplémentaires. Ils plongent, tournent sur eux-mêmes, grimpent sur leur rocher et se précipitent dans le bassin."
6. "Quand elle n'est pas effrayée par la foule, elle sort de l'eau à son appel, elle arrive en boitant sur les moignons qui lui servent de pattes, lui prend à la

main le poisson qu'il lui tend, et même, en récompense, l'embrasse délicatement sur la joue."
7. "dernier portrait de Neff peu de temps avant sa mort."
8. "Il se livre alors avec son gardien à des amusements variés; sur un signe de lui, il élève ses deux pieds de devant sur un gros tronc d'arbre, et se dresse de toute sa hauteur, en rejetant sa trompe en arrière."
9. "L'Eléphant Kouch fait le beau."
10. "Eléphant d'Afrique Saïd qui a étouffé son Cornac Neff le 30 Août 1905."
11. "L'éléphant est un animal très intelligent et très doux; il est vindicatif; mais dans le cas actuel il est certain qu'il n'avait pas à se venger de son cornac qui l'aimait et le soignait avec dévouement. Comment donc expliquer ce brusque accès de fureur? Je crois qu'on peut l'attribuer tout simplement à la captivité. Qu'on se représente ce qui doit se passer dans l'esprit d'un animal intelligent qui depuis de longues années se voit enfermé dans un espace restreint . . . . Un accès de nervosité est bien excusable." "l'état hygiénique déplorable"; "une institution traditionelle qui ne sert pour le moment qu'au délassement de la population."
12. "assez vilain caractère"; "Il ne faut pas lui en vouloir: ses dents ne sont pas fameuses et rien n'aigrit comme les maux de dents. On a dû extirper à Saïb [sic] jusqu'à son dernier bout de défense et son gardien, dentiste improvisé, est obligé de surveiller de près sa mâchoire."
13. "sont trop malheureux pour n'être pas excusables de ces moments de fureur"; "éléphant assassin"; "Saïd a succombé aux cuisants remords que lui causait son crime"; "on l'a laissé méditer en paix, dans le régime cellulaire auquel il était soumis depuis longtemps."
14. "qu'il n'est en rien la cause."
15. Kako, dit le Terrible"; "Liza, sa compagne."
16. "instinct de brute le poussa à supprimer l'obstacle qu'il redoutait."
17. "crime de l'hippopotame."
18. "A l'exemple des criminels célèbres, Kako mérite d'être biographié." "un personnage important."
19. *Kako le terrible, D'après un véritable fait divers qui eut lieu au Jardin des Plantes de Paris en 1903.*
20. "retrouver des vies animales, restituer leur existence, montrer leur singularité."
21. "en émettant de petits grognements"; "le terrible"; "animé d'un instinct de brute qui finalement ne l'avait peut-être jamais quitté."
22. "Les collages avec les vieilles cartes postales permettent une sorte de mise en abîme puisqu'à l'intérieur de mon hippopotame se trouve la photo d'un autre hippopotame."
23. "n'ont pas du tout le sens du danger, ils semblent avoir oublié qu'un animal sauvage reste un animal sauvage, même en captivité."

References

Baratay, Éric (2017). *Biographies animales. Des vies retrouvées.* Paris: Éditions du Seuil.

"Broyé par un hippopotame" (1903). *Le Petit Journal*, July 2.

Cartes postales du Muséum national d'histoire naturelle, Ms 3315 and 3316. National Museum of Natural History Archives, Paris, France.

"Colère d'hippopotame" (1901). *Le Petit Journal*, September 17.

Colley, Ann C. (2014). *Wild Animal Skins in Victorian Britain: Zoos, Collections, Portraits, and Maps.* Burlington, VT: Ashgate.

"Dans les coulisses du Jardin des plantes" (1900). *Lectures pour tous*, August.
"Echos" (1906). *La Justice*, January 25.
"Éléphant meurtrier" (1905). *La Presse*, August 31.
Estebanez, Jean (2010). "Ceux qui sont proches: les soigneurs au zoo." *Sociétés*, no. 108: 47–57. https://doi.org/10.3917/soc.108.0047.
Guyot-Daubès (1885). "Les Phoques savants." *La Nature*, October 24.
Kreger, Michael D., and Mench, Joy A. (1995). "Visitor-Animal Interactions at the Zoo." *Anthrozoös* 8, no. 3: 143–58.
Laloy, L. (1905). "À propos de l'éléphant du Jardin des plantes de Paris." *Le Naturaliste*, September 13.
"La Victime de l'hippopotame" (1901). *Le Petit Journal*, September 23.
Le Curieux. "À travers Paris" (1903). *Gil Blas*, July 6.
"Les Méfaits d'un éléphant" (1905). *Le Radical*, September 1.
"L'Hippopotame meurtrier" (1903). *Le Petit Journal*, July 27.
Milne-Edwards, A. (1891). *La Ménagerie. Rapport au ministre de l'Instruction publique*. Paris: G. Masson.
"Obsèques du gardien Lancelle" (1903). *La Lanterne*, July 7.
Petre, Dominique (2016). "L'Histoire d'un infâme hippopotame." *Ricochet*, December 14. www.ricochet-jeunes.org/articles/lhistoire-dun-infame-hippopotame.
Polack, Emmanuelle, and Barroux (2013). *Kako le terrible*. Geneva: Éditions La Joie de lire.
R. (1904). "Un criminel en gélatine." *Le Pêle-mêle*, March 20.
Raveret-Wattel, C. (1882). "Rapport au nom de la commission des récompenses." *Bulletin de la Société nationale d'acclimatation de France* 9: XCII–CXI.
Ritvo, Harriet (1996). "The Order of Nature: Constructing the Collections of Victorian Zoos." In *New Worlds, New Animals: From Menagerie to Zoological Park in the Nineteenth Century*, edited by R.J. Hoage and William A. Deiss, 43–50. Baltimore: Johns Hopkins University Press.
Tout-Paris (1907). "Bloc-notes parisien. La mort de Saïd." *Le Gaulois*, January 31.
"Tué par un éléphant" (1905). *Le Temps*, September 1.
"Tué par un hippopotame" (1903). *La Croix*, July 3.
Un Dimanche au Jardin des plantes (1910). Paris: Librairie Larousse.
"Un drame au Jardin des plantes" (1905). *Le Journal*, August 31.
"Un éléphant assassin" (1905). *Nouvelliste valaisan*, September 5.
"Un éléphant qui tue son gardien" (1905). *L'Illustration*, September 2.
Wood, Amy Louise (2012). "'Killing the Elephant': Murderous Beasts and the Thrill of Retribution, 1885–1930." *The Journal of the Gilded Age and Progressive Era* 11, no. 3: 405–44.

8 "Seeing Eye to Eye, Through a Glass Clearly"? Blurring the Boundary Between Humans and Animals

Quentin Montagne

In *The Trouble with Bubbles*, a short story written in 1953, Philip K. Dick imagines the commercialization of a singular artefact. With the expanded use of robots, humanity is free from all labour, and their attention turns to the heavens in the hope of discovering a new form of life in the solar system. Following numerous space expeditions, it unfortunately proves to be empty. The dream shattered, a deep depression spreads over the idle population. At this point, the Worldcraft Industries company offers a new hobby; since mankind is incapable of finding another inhabited world, they can now buy their own. At a sub-atomic level, scientists have discovered how to trap a multitude of microscopic planets inside transparent bubbles, and each person can control the development of their own planet. In a short time, this product is distributed. Almost everyone gets to own a bubble, and competitions are organized, with the most "advanced" planet winning. Despite the notable results, the end of the event is marked by the destructions of all the glass balls. In a collective movement and cry of joy, each participant brutally smashes his toy-world.

The aquarium is extremely similar to the spheres imagined by Philip K. Dick. Alongside the formal analogy between Worldcraft's product and traditional red fish bowl, the latter became a revolutionary hobby when Robert Warington launched its commercialization in 1850. After the short and intense trend of *aquariomania* (Lorenzi 2009; Olalquiaga 2008, 45–53), this object has grown in popularity since the beginning of the 1980s. Alongside of the construction of public aquariums, more amateurs indulge in the hobby at home every day, as numerous competitions, national or international, reward the most skilled creations.[1] It is also worthy of note that most of the current competitors also unmake their creations once a prize is awarded in order to create new ones, though without the hysteria described in the story. However, the most significant resemblance in our analogy is in the link established between the observer (owner or spectator) and the encapsulated world. Similar to Philip K. Dick's bubbles, it is through its enveloping screen that the aquarium is defined. It offers an intense proximity with a foreign environment, while still maintaining a very fine frontier, transparent but still tangible.

Through this screen, both the aquarium hobbyist's design and our gaze are structured. As a true scenographer giving free rein to his imagination, the hobbyist decides on the setting, chooses its actors, and goes on to build the stage of the show. The tank encapsulates a fake world, visible from the other side of the glass.

1. *Zoodram*

Zoodram are artworks by Pierre Huyghe that express the specific posture of the scenographer. The shapes and content of the water tanks of these artworks do not really diverge from those presented in public aquariums or some imposing home creations. The artist's retrospective exhibition of 2013 at the Centre Pompidou includes one of these artworks entitled *Zoodram 4*. Set on a simple wooden plinth, the tank of approximately a thousand litres hosts a crustacean population amongst an apparent chaos of live rocks. This scenery is illuminated from above with a reddish glare diffused from a suspended light-box. No other aquarium equipment is to be seen, in order to attract the spectator's gaze to this enclosed undersea universe. The accessories that are necessary for maintenance—the filters, the pumps, and the secondary tank—are hidden within the device. Inside, the lead role is taken by a hermit crab. Carrying a copy of *The Sleeping Muse* by Brancusi for shelter,[2] it strolls with difficulty amongst the rubble. Often overlooked by critics, a small group of yellowline arrow crabs offer a genuine contrast with the hermit crab. Discrete and unobtrusive with their plain colour, their abnormally slow movement suggests a dance. Equivalent to slow motion, their long and thin legs carefully explore each crevice of the bedrock in search for food.

According to Pierre Huyghe, this aquarium seals an autonomous ecosystem, and no further interaction is necessary. "A collection of animals, chosen specifically, are placed under certain conditions. They do not act; they have no other behavior than what is natural to them" (Huyghe 2011). Even if there appears to be no scenario or drama in this mind-set, the artist still maintains some control over the occupants. Along with the confinement that ineluctably modifies the action of the captives, this artwork benefits and takes advantage of the natural behaviour of the animal. Already in his short film *The Hermit Crab* of 1927,[3] Jean Painlevé altered the behavioural pattern of the animal: we perceive a crustacean miniature football player through the filmmaker's point of view as it instinctively turns a cork ball around in search of a potential cavity to be used as shelter.

Though the artist prefers to qualify his artworks as a "situation" more than a "stage" (Huyghe 2011), an aquarium establishes the same context as a theatre in which the aquarist can be compared to a stage director. The hermit crab is directed, as it has no other choice in *Zoodram 4* than to inhabit a copy of *The Sleeping Muse*, a costume imposed by the artist. The orchestrated situation within the tank is determined by its glass boundaries and therefore becomes a representation.

Just like the original design of Warington (Warington 1852, 5–7) and most available aquariums on the market today, *Zoodram 4* has a parallelepiped shape. Positioned in the middle of the room, its sculptural aspect is highlighted. The observer can walk around the artwork without restriction, embracing different depictions on the successive vertical glass panels. A thin black border drawn by silicone sealant gives the impression that the glass is framed. In the same way as a window, an aquarium opens "to a scene different from the one where the public is situated" (Seux 2014, 17). The term "living picture" is often employed to qualify aquariums, and it is fully applicable here. If this device draws heavily upon the concept of painting—according to the definition given by Alberti in 1435 (Alberti 1992)—it primarily introduces elements of Italian theatre from the end of the sixteenth century onwards, built to conform itself to a pictorial style and the rules of perspective, in effect during Cinquecento (Klein 1970, 294–309).

Every aquarium is a scene-painting. It exhibits an illusionary and unified setting, built to accommodate the external and dissociated position of the viewer. The diaphanous and glazed panels of the aquarium meet "the condition of visibility of the represented world" (Marin 1994, 305). It materializes the theatre's invisible screen. The frame is an "indispensable auxiliary of representation" (Hénin 2004, 24) and is made possible by the structure of the tank itself. Aquariums are containers that aim to capture and concentrate the viewer's gaze, spontaneously excluding all other objects from the line of sight. The aquarium "is not simply given to be seen among other objects" (Marin 1994, 347) but is in itself an object of contemplation. As such, the aquarium might be unadorned, like Pierre Huyghe's artwork, or it could be sumptuously ornamented. In both cases, the container blends in to enhance its content; it "acquires the exact same rhetorical status as an ornament that must emphasize something without drawing attention [to itself]" (Hénin 2004, 44).

Inside the theatre of an aquarium, the animal becomes an image. To the exception of zoological park mascots, specimens presented in a tank are only perceived as representatives of their species. The descriptive labels on the fringe of the containers are unambiguous, and their information is generally limited to a brief description of the species represented by one or several replaceable entities in captivity. Instead of creating a sense of proximity with the animal, this strengthens the objectification of the ancient aquatic fauna.[4] Unlike terrestrial animals, these bizarre living beings generally fail to create empathy and recognition. Except for the unique case of the axolotl, anthropomorphism is made impossible. The spectator is physically separated from the animal and can therefore only establish an indirect relation with it. This distance created by the aquarium as an object has been problematic to aquarists from the very beginning.[5] Since the early twentieth century, aquarists have never ceased to try to break the frontier between man and the subaquatic beings inside the aquarium.

2. Picturing the Encounter

In the obscurity of public aquariums, the similarity between fish tanks and a stage is intensified. The entrance of the Cinéaqua in Paris displays freshwater tanks that still let daylight in. Beyond this point, the visitor is plunged into dark spaces, only lit by glowing screens formed by huge reservoirs built into the walls. The rather plain environment seems to fade out of the spectator's field of view, as he is captivated by the aquatic beings. Spectacularity is reinforced later in the visit. Lines of armchairs are set in front of some glass panels, and, in the heart of the structure, there is a concert hall, complete with speakers, coloured spotlights, and a disco mirror-ball. Like Valencia's Oceanogràfic auditorium and San Sebastian's modest aquarium, hanging curtains frame the scenery of rocks and shoals of fish. These are explicit stage elements.

The similarity between aquariums and theatres is not only comparable in shape and structure. In the first years after Warington's invention, public aquarium managers tried to conceive underwater animations and shows in their facilities. Yet these attempts faced an obstacle: acts in the tanks lacked singularity, intrigue, and drama. Aquariums offer unities of action, time, and place, and thus the settings cannot really evolve. Nowadays, the occasional video projection can only distract from this to a certain point. Furthermore, the behaviour of captive animals, as observed by Pierre Huyghe, remains predictable, repetitive, and monotonous. Just like a red fish in a jar, whale sharks of the Kaiyukan Aquarium in Osaka circle endlessly around their huge tank, whereas a green turtle of the Vannes Aquarium rubs pathetically against the border of its small pond.

To conceive of a scenario in the midst of the perpetual ballet of fish and mollusks, aquarium specialists have no other choice than to add actors or mechanical devices to the scenery. The Aquarium of Paris is a good example of these add-ons. It was conceived as a temporary facility, built for the Universal Exhibition of 1900. In one of its water tanks, called *The submarine volcano*, visitors could discover an artificial eruption of gas and smoke in a cluster of basaltic rocks inhabited by a population of sea breams, congers, turbots, and red mullets (Image 8.1). In their desire to illustrate the latest geological discoveries of the time, especially after Krakatoa's volcanic eruption, the Guillaume brothers, in charge of the Aquarium, imagined a device which was to show

> the luminous effect produced by gases . . . escaping from an underwater volcanic crevice. . . . From time to time, the rock seems to open, a red glimmer appears; and, from this miniature volcano, an column of flame escapes and rises into shiny large globules that are pushed towards the surface.
>
> (Guide-Souvenir de l'Aquarium de Paris, 48–49)

Image 8.1 "The Submarine Volcano," in *Guide-Souvenir de l'Aquarium de Paris*, Paris, Simonis-Empis, 1901, p. 47.

A few metres away stands *The Coral Sea* (Image 8.2). Among the madreporians and tridacnas flourish an impressive fauna of gurnards, flying gurnards, oysters, and Cottidae, with several species of shark: sawfish, rays, hammerhead, and tiger. In this crowded environment, "the promoters of the Aquarium wanted to display to their visitors an exact reproduction of this curious spectacle" (Guide-Souvenir de l'Aquarium de Paris, 56), namely of the Sri Lanka's rayed pearl oyster fishermen. "Divers with exceptional skills—one of them, like a true amphibian, stays four minutes at the bottom of the water—come every day [. . .] to play the role of fishermen from the Manaar and Ceylon Islands in the coral tank" (Guide-Souvenir de l'Aquarium de Paris, 56). These figures had been popularized

Image 8.2 "Mermaids inside the coral tank," in *Guide-Souvenir de l'Aquarium de Paris*, Paris, Simonis-Empis, 1901, p. 51.

at the time by Jules Verne's famous novel, *Twenty Thousand Leagues Under the Sea* (Verne, 221–234).

Fable and reality unite, with "long-haired mermaids interlaced with floating algae [cutting across the bottom of the aquarium,] that seem to stir within the waves" (Guide-Souvenir de l'Aquarium de Paris, 56). The visitor's attention to the aquatic world is encouraged by the divers observable inside the aquarium. Face to face with a fellow human being immersed in an apparently hostile environment, the spectator imagines himself in his place. By the simple presence of an individual in the middle of shoals of fish and other animals, these species seem more familiar than what they appeared to be at first glance. These shows have become widespread in contemporary aquariums, and what used to be a quirk is now the main attraction. One of Tokyo's Sunshine Aquarium performance involves a diver in the midst of sharks and reef fish, with a voice-over describing the fish as she catches and caresses them delicately.

Public aquariums have also incorporated another type of animal show, *dolphinariums*, common in zoological parks since the nineteenth century, with roots going back to the tradition of animal-taming and circuses. The father of Carl Hagenbeck was a pioneer with his seal show at Hamburg's fair in 1858 (Baratay and Hardouin-Fugier 1998, 128), and similar entertainment can now be seen at Paris's zoological park and at the Sunshine Aquarium. Under the direction of a breeder or trainer, animals perform a series of figures and tricks for food, either on podiums or cutting across specifically designed platforms (Image 8.3). Although these shows are part

Image 8.3 "In the London Zoo, spotlights are set up: the seal comes to inspect the work" (press photograph).

Source: Press agency Mondial Photo-Presse, Paris, 1932. Black-and-white photograph, 13 × 18 cm.

Paris, National Library of France, Prints and Photography Department.

of the public aquarium's entertainment, they are very different from those in the aquarium-container. Their purpose is not to present an animal in its natural environment, or to display fantasized subaquatic scenes, but rather to highlight man's interaction with the animal; because the taming of fish and molluscs is impossible, such shows only include birds and mammals. Whatever jumps made by dolphins or poses made by orcas, the encounter between man (trainer and spectator alike) and animal always occurs above water, in the open air. Touch tanks and touching-basins also allow visitors to have physical contact with fish and shellfish in shallow water, and it is rare to find a public aquarium today without these child-friendly installations. As at the Georgia Aquarium, where individual encounters with dolphins and penguins are organized, the visitor only gets his arm or hand slightly wet while remaining in his own dry environment.

3. The Spectator's Immersion

The shows actively engage man and aquatic animals into a relationship, but such contact can only be staged for the public, and the spectator does not connect with the animal. Therefore, aquarium managers have developed other immersive techniques to overcome the separation created by a frontal scenography approach. Once again, the Aquarium of Paris in 1900 is a good example. Its two main water tanks called *The Sunken Ships* seem to be united as one (Image 8.4). The first is over 70 cubic metres of water and contains the remains of the *Graf von Bismarck-Bremen* sailboat,

Image 8.4 " The Sunken Ships," *Guide-Souvenir de l'Aquarium de Paris*, Paris, Simonis-Empis, 1901, p. 34–35.

which sunk in transit for Cherbourg. Next to it is "the formidable bow of the steamboat that approached the sailboat out of the harbor and sank with it" (Guide-Souvenir de l'Aquarium of Paris, 38). Its hull is used as a junction wall with the other fish container. Among the remains of the shipwrecks, the public can observe a multitude of algae, madreporians, molluscs, fish, sharks, and turtles. The sunken sailboat is used as an area for helmet diving.

> At certain hours of the day, a wave oscillates in the distance of the container; the frightened fish flee in all directions, and crabs hide in their holes. Suddenly, divers appear, their copper helmets reflecting a tawny color from the muted lights of their lanterns. They cut across the rocks and the ship's debris before breaching the shipwreck, and they soon come back out carrying the cargo away.
>
> (Guide-Souvenir de l'Aquarium de Paris, 40)

In front of glass panels three and a half metres tall, visitors experience this aquatic show at ground level, in the same settings as in the tanks. The bow of the *Graf von Bismarck-Bremen* bridges through the large glass screen and into the public elliptical room of the Aquarium. The second bordering tank is built into the walls and coated with rocks

> on which plants and sea creatures like to attach themselves; near a longline of mussels, yellow petal-like tentacles of sea anemones fan out, and, here and there, appear to contrast the darker shades of the rock, the shiny shells of common wentletrap, turbonilla, and pearl oysters.
>
> (Guide-Souvenir de l'Aquarium de Paris, 32)

From a gallery of living paintings (Image 8.5), the public aquarium becomes an entire environment.

The aquarium built for the International Exhibition of 1867 had its underground walls and ceiling fully glazed, enveloping its visitors inside an aquatic cavern scenery (Image 8.6). The Guillaume brothers followed this model for their attraction, wanting to give to the public an impression of being

> suddenly transported into an underwater clearing. . . . To achieve this, they stretched a huge *velarium* over the entire room. Behind it, they arranged a set of powerful projector-lights on circular tanks, filled with sea water and marine plants. It was also inhabited by fish, shellfish and crustaceans. This layout exceeded expectations. The ceiling created rays of constantly agitated light, with the entanglements of vivid algae and gigantic shadows of monstrous fish and fantastic decapods. This is not a minor achievement in the overall success of the Aquarium.
>
> (Guide-Souvenir de l'Aquarium de Paris, 34)

Image 8.5 View of the aquarium in the Jardin d'Acclimation of Paris, engraving by Bertrand, in *Le Monde illustré*, 10 January 1863.

Image 8.6 View of the Marine Aquarium, Paris International Exhibition, 1867, by Cotton-Smeeton, in *L'Illustration. Journal universel*, 19 October 1867.

The spectator is no longer solely entertained by scene-paintings arranged behind panels but is included in an immersive show and is physically engaged. He even tends to forget the presence of these invisible borders.

The illusion of a widespread space, no longer enclosed behind glass walls, is connected to the tradition of panoramas. This device was developed by Robert Barker in 1787 and popular in the following century, especially in the context of International Exhibitions. Germain Bapst defines the panorama as

> a circular painting displayed in such a way that the eye of the beholder is placed at the center of an all-embracing view, made of canvas. . . . To establish the illusion, it is necessary that the eye always meets figurative elements, in the correct proportions and exact tones of color, and this, on whatever point it is drawn towards. Nowhere must be found the view of real objects that could serve as a comparison; so that, while he is only looking at a work of art, he thinks he is in the presence of nature.
>
> (Bapst 1891, 8)

At first, this device is exclusively pictorial art. Canvases are spread out on the inside walls of purpose-built circular rotundas. An elevated observation-platform designed to host the public is set in its centre, keeping them at a certain distance from the painting. The platform has a top cover that conceals the semi-glazed ceiling of the rotunda, the source of the natural light flooding in. Fabric of matching colour is stretched from the platform's edge to the canvas paintings, hiding its borders.

Barker's invention was well-received and further developed. Its dimensions increased considerably, and for the first time in 1830, Jean-Charles Langlois added three-dimensional decorations and objects as a foreground. One of his most famous artworks is *The Battle of Navarino*, the design of which served as a forerunner to the Aquarium of Paris. Unlike previous attractions that were only visual and created a passive attitude, Langlois put the spectator at the centre of the action:

> Knowing that *The Scorpion* was a seventy-four gun ship that had been particularly successful in battle, and that it had recently been put up for sale, . . . he became the acquirer of the cabin and poopdeck of this vessel and used it as the platform for the spectators. His panorama was arranged so that spectators would enter a dark passageway that led through the inner corridors of the platform, transformed into a ship's interior. First entering the gun deck of 18, the spectator finds himself in the officers' room. . . . [T]hen, climbing up a small staircase, he arrives in the commander's dining room, entirely furnished, with its chairs, chest of drawers, telescope, compass, various instruments, and cockpit tables. He climbs up to the second floor,

> and from the commander's quarter he walks to the quarterdeck; from this point he can see the fight around him. A Turkish fireship approaches the *Scorpion* ahead, propagating to fire; sailors work at pulling their ship out of the way; smaller boats have tied the fireboat up and attempt pulling it away; a Turkish warship blows apart at a short distance; farther, the *Armide*, a French frigate, tows an English corvette and takes its place, crushing a Turkish frigate called *The Belle Sultane*. The English sailors, perched on the heights of the corvette, cheer the *Armide*'s crew. In the foreground on the right side of *The Scorpion*, a Turkish schooner is about to sink, scattering debris of masts, warships remains, and men.
>
> (Bapst 1891, 23–4)

In the same way as *The Battle of Navarino*, the visitor of the Aquarium of Paris has the illusion of participating in the show. The background framework is still there but now disregards the separation between the underwater microcosm and public space. The barrier now blends into the scenography, becoming a crucial linking element to a "vast *trompe-l'oeil* in which the whole room is drawn in" (Hénin 2004, 35).

Public aquariums enclose worlds of their own. In the same way as panoramas, they provide *heterotopia* inside their walls. The representation starts right at the entrance of the building. Barker's invention had the concept of a subterranean entrance to his rotunda already built-in. Walking through the dark corridors, the visitor was disoriented, losing track of daylight and time, which created the perfect conditions to appreciate the luminous scenery-painting. The Guillaume brothers opted for the same method to break with reality, keeping the architecture's size out of vision and surprising the public with the huge size of their aquarium. The illusion is the same at the Marine Aquarium of the International Exhibition of 1867, at Berlin's Unter den Linden Aquarium, and recently at the Tropical Aquarium of the Palais de la Porte Dorée in Paris. With hardly any exceptions today, public aquariums set a dark passageway before giving access to lit tanks.

The entrance of the Aquarium of Vannes and the current Vincennes Zoo adopt the appearances of cavities dug into the rock. This choice creates a symbolic transition into the belly of the earth, especially as its rocky walls extend inside. To enter the visit, the Kaiyukan Aquarium has long sets of stairs and complicated corridors leading to the *Aqua Gate*, a symbolic entrance to the exhibition space. Its bright-blue water tank forms an archway and is inhabited by a large number of eagle rays and orange fish. The entrance to the *Cité de la Mer* in Cherbourg is also situated on higher levels and the circuit then takes visitors downwards, through darkened spaces amongst showcases, models, machines, and video projections, until it reaches the area occupied by the aquariums. A small warning sign before this entrance states, "Chez eux, nous ne sommes pas chez nous" ("Here they dwell and we do not"). From this threshold, the public breaches into another world, a foreign environment of fish, crustaceans, and other aquatic organisms.

Like the *Cité de la Mer*, the Grand Aquarium of Saint-Malo gives access to the exhibition on the first floor. Behind the ticket counters, visitors climb up some stairs decorated by wooden crates, all in dim light. As if echoing *The Battle of Navarino*, they arrive in a captain's great cabin, and from an elevated platform, they discover a nineteenth-century cabinet replica containing a desk globe, maps, ancient zoological essays, and a collection of natural history specimens. While the Aquarium of Paris and Langlois's panorama present one scenography alone, the Aquarium of Saint-Malo, just like the Kaiyukan in Osaka, offers a series of different rooms to the visitor. Alongside a rockwork scenography and a shipwreck, similar to those of the Guillaume brothers, certain areas of the edifice offer a very unique type of immersion to the public. Instead of a classic underwater cavern, the public sets foot on a scientific submarine deck and then an underwater architecture base. In this scenario, the glass barrier has no drawbacks, and even reinforces the design, reworked into underwater portholes bordered with iron and copper, and assembled with bolts and nuts. Once again, the spectator is in the centre of the action, but still separated and protected from the water.

The *Nautibus* is an attraction at the end of the visit that proposes an almost tangible fiction. Entering through a passage themed with rusty metal panels, switches, and push buttons, the public is invited into a small capsule diving to the depths of the sea. The airlock hatch at the surface opens to narrow, spiral stairs. Inside, three metallic stools are set in front of a translucent screen. The descent on guided rails begins, though, in reality, the immersion is closer to an amusement park ride than an authentic submersion. The device is clearly inspired by Daguerre and Bouton's Diorama, a rotating auditorium to carry the comfortably sitting visitor to different picturesque scenes, creating realistic window illusions with the use of light. Contrary to the Diorama in Sanson Street, the illusion of the *Nautibus* ends at the first window stop. As the visitor crew discovers Russian sturgeon and moray eels outside of the pod, the glass barrier is unfortunately noticed. What turns out to be a tank finally reveals the scenography's construction.

The *Ring of the Seas* is more convincing (Image 8.7). This platform is placed within a 600,000-litre tank inhabited by turtles, dusky groupers, and sharks. Despite their 20-centimetre thickness, the five curved panels composing its surface are connected without any apparent joint. Like a panorama, the screen offers a 360-degree scene, the huge transparent cylinder exceeding the dimensions of the room to hide its lower and upper edges. This overwhelming tank takes full advantage of technological improvements made in marine exhibits in the second part of the twentieth century. Instead of using glass, which becomes less translucent, heavy, and

Image 8.7 "The Ring of the Seas," Saint-Malo, Grand Aquarium of Saint-Malo, France.

Source: Quentin Montagne.

brittle in thick layers, the architects make use of polymethylmethacrylate (PMMA). This synthetic material was used in public aquariums for the first time at Florida's Marineland Park in 1960. PMMA is lighter, more resistant, and has a greater transparency than glass, and makes curved, spherical, and polyhedral container-shapes uncomplicated to produce.

In 1979, the San Francisco Steinhart Aquarium still used metal plates to secure the glass, and the Aqua Dome of the Seattle Aquarium, inaugurated the same year, used reinforced concrete structures to seal the immense glass panels. It was only at the end of the 1980s that new improvements in the development of immersive devices for public aquariums were made. In 1989, the Tokyo Sea Life Park inaugurated a long, curved Perspex screen with no reinforced joints, as it discharged from the frame. The distance maintained in between the public and the tank creatures was thus thinned and refined. Continuity no longer resided in homogenous scenery between the aquarium tanks, as their shape was sufficient to lead the public through an immersive experience. What used to enclose a frontal view now offered a whole new vision. In 1998, the Discovery Cove in the New York Aquarium created fully transparent tunnels, giving the impression of "standing right under a cresting wave" (Taylor 1993, 34), and this equipment has since been used in many public aquariums. The visitors of the Shanghai Ocean Aquarium enter an acrylic glass tunnel from the top of a gigantic reservoir and descend below the surface on an escalator, and since 2003, the customers of the Radisson Hotel Berlin have used an elevator to descend down the AquaDom, an impressive 25 metres tall cylindrical tank. Many public aquariums design their exhibitions around round-crowned tunnels, like the Monterey bay aquarium, and in Dubai's Atlantis water park, visitors can even ride entire translucent-tube slides surrounded by sharks and other fish.

It now seems that the last way to completely erase the frontier for the public would be to enter the other side of the glass. This has been the Sub Sea System company's ambition since 1999 with their recreational diving equipment called the *Sea Trek Diving System*. This semi-autonomous suit uses a simple hermetic diving helmet supplied with air through pipes that reach the surface. First used on the beach for tourists, this equipment was also acquired shortly after release by public aquariums such as the Manchester Sea Life, the Cancun Aquarium, and the S.E.A. Singapore Aquarium. At OdySea in Scottsdale, Arizona, the public is invited to enter the main reservoir and walk "among thousands of curious and colorful marine creatures of Indo-Pacific regions" (OdySea Aquarium n.d.). The tank in this case is entirely interactive, and what used to be a frontier now offers the possibility of direct contact with the animal. Without the use of Sub Sea System's equipment, many public aquariums still offer dive sessions inside their tanks. For $189.95 ($233.95 for non-members), the Georgia Aquarium allows visitors to swim among whale sharks, while at Florida Keys Aquarium Encounters, visitors dive with a snorkel among reef fish fauna. While London's SEA Life or New York's Long Island Aquarium keep their visitors immersed behind a cage or a net, the Denver Downtown Aquarium and Cheshire Blue Planet leave the public to swim freely among sharks.

Penetrating the scenic cube of the aquarium, the visitor-diver negates the distance that separates him from the animal. He leaves his spectator position to become an actor in the same manner as the sea creatures. For the public on the other side of the glass, he takes part in the active picture, becoming one character among others. Even though the educational, scientific, and entertainment values driving these underwater experiments are questionable, it is the simple act of entering the tank that destroys the illusion for the diver. Through the scenography, he distinctly sees all the parameters of the tank, and his encounter with the animal kingdom remains incomplete. Despite the evolution of underwater exploration devices, the descent is limited by the supply of oxygen. Man remains a temporary visitor under the surface of the water. Terrestrial and underwater worlds are inexorably strangers to each other. Jacques Rougerie reminds us that the human eye "is not designed for underwater vision. It cannot clearly see in an environment 800 times denser than air, with very different optical properties, the main one being the change in the refractive index. The diver uses a diving mask, the only technique he can use to restore balance" (Rougerie, 79–80). In nature or in the tank, perception of the underwater environment remains inexorably regulated and filtered. In the end, the diving mask simply replaces the aquarium screen, and, just like Warington's invention, photography, or video, it encloses and informs our gaze.

Notes

1. As an extremely popular hobby, many regional, national, and international competitions are organized around aquariums. Some events are exclusively dedicated to the beauty of aquatic specimens, through long breeding processes. Like an agricultural fair, the Aquarama International Fair includes more than a dozen awards: International Goldfish Competition, International Dragon Fish Competition, Crystal Bee Shrimp Competition, and so on. Other competitions are dedicated to the decoration and composition of water tanks. Different categories, defined by size, volume, or contents of the tanks, are rewarded differently depending on the organizers. It would be impossible to list all these competitions; almost every country in Asia and Europe hosts regional and national championships. The most famous ones are the Aquatic Scapers Europe, the Acuavida Aquascaping Contest, the AGA Aquascaping Contest, the Aquascaping World Competition and, the most prestigious of all, the International Aquatic Plants Layout Contest.
2. Constantin Brancusi, *The Sleeping Muse*, 1910. Polished Bronze, 16 × 27, 3 × 18, 5 cm. Paris, Centre Pompidou.
3. Jean Painlevé, *The Hermit Crab*, 1927. Black-and-white film with sound, 13′11″.
4. This particular status of aquatic fauna is particularly noted by Éric Baratay in two of his works (Baratay 2003, 2007). Unlike mammals, notions of intelligence and sensitivity are still largely denied to aquatic fauna. It was not until 2003 that researchers from the University of Edinburgh were able to demonstrate the state of physical suffering of a rainbow trout for the first time (Schiermeier 2003). This did not prevent the ECNH for "Federal Ethics Committee on Non-Human Biotechnology" (Switzerland) from qualifying the sensitivity of fish a decade later (Le Temps 2014; CNEH 2014).
5. According to Stephen Jay Gould, contrary to ancient containing systems, fish ponds, and basins, the aquarium allows us to see the aquatic world "not from above through the opacity of flowing waters with surface ripples, but eye-to-eye and from the side through transparent glass and clear water" (Gould 2001, 68).

References

Alberti, Leon Battista (1992). *De la peinture*. Paris: Macula.

Bapst, Germain (1891). *Essai sur l'histoire des panoramas et des dioramas. Extrait des rapports du jury international de l'Exposition universelle de 1889*. Paris: Imprimerie nationale.

Baratay, Éric (2003). *Et l'Homme créa l'animal*. Paris: Odile Jacob.

Baratay, Éric (2007). *Les Planches du Dictionnaire universel d'histoire naturelle de Charles d'Orbigny: portraits d'animaux*. Lyon: Fage.

Baratay, Éric, and Hardouin-Fugier, Elisabeth (1998). *Zoos: histoire des jardins zoologiques en Occident, XVIe-XXe siècles*. Paris: la Découverte.

CNEH (2014). *Utilisation éthique des poissons*. www.ekah.admin.ch/fileadmin/ekah-dateien/dokumentation/publikationen/f-Broschuere_Ethischer_Umgang_mit_Fischen.pdf.

Dick, Philip K. (1953). "The Trouble With Bubbles." *World of If* (September): 189–214.

Gould, Stephen Jay (2001). "Se voir face à face, clairement, à travers un verre." In *Les Coquillages de Léonard*, 66–83. Paris: Le Seuil.

Guide-souvenir de l'Aquarium de Paris. Exposition universelle de 1900 (1901). Paris: H. Simonis- Empis.

Hénin, Emmanuelle (2004). *La Scène comme tableau*. Poitiers: La Licorne.

Huyghe, Pierre (2011). *Interview par Timothée Chaillou*. www.timotheechaillou.com/conversations/pierre-huyghe/.

Klein, Robert (1970). *La Forme et l'Intelligible*. Paris: Gallimard.

Le Temps (2014). "La Souffrance des poissons reconnues." Accessed 10 January 2018. www.letemps.ch/sciences/2014/12/01/souffrance-poissons-reconnue.

Lorenzi, Camille (2009). "L'Engouement pour l'aquarium en France (1855–1870)." *Sociétés & Représentations*, no. 28: 253–71.

Marin, Louis (1994). *De la représentation*. Paris: Gallimard.

OdySea Aquarium (n.d.). "SeaTREK Helmet Diving." Accessed 5 January 2018. www.odyseaaquarium.com/seatrek-helmet-diving/.

Olalquiaga, Céleste (2008). *Royaume de l'artifice. L'Émergence du kitsch au XIXe siècle*. Lyon: Fage.

Rougerie, Jacques (1978). *Habiter la mer*. Paris: Éditions maritimes et d'outre-mer.

Schiermeier, Quirin (2003). "Trout Study May Fuel Anti-Angling Debate." Accessed February 5, 2014. www.nature.com/news/2003/030430/full/news030428-9.html.

Seux, Christine (2014). "Écran(s)." *Le Télémaque*, no. 45: 15–25.

Taylor, Leighton (1993). *Aquarium: Windows to Nature*. New York: Prentice Hall General Reference.

Verne, Jules (1871). *Vingt mille lieues sous les mers*. Paris: Hetzel.

Warington, Robert (1852). "The Aquatic Plant Case, or Parlour Aquarium." In *The Garden Companion, and Florists' Guide*, edited by Arthur Henfrey, 5–7. London: William S. Orr and Co.

9 Flesh, Fur, and Forgetting

Kari Weil

The terms "creature" and "creaturely" have become newly popular in the field of animal studies. On one level these terms remind us of the theological context of creation in which we are all God's creatures. Whether human or non-human animal, we share our role as created beings rather than as creators. The creaturely is thus associated with a kind of passivity that is also at play in recent, post-secular uses of the term that call attention to the shared embodiment of humans and other animals. Anat Pick (2011) calls for a poetics that attends to the beauty of the flesh and of our shared vulnerability as physical and mortal beings over and above our powers of mind. Eric Santner has focused especially on the biopolitical dimension of passivity as an "exposure to excitations of power," one that, however, he sees as peculiar to human life. Thus he speaks of the "essential disruption that renders man creaturely" (Santner 2006, 12). Creaturely, in this regard, is life shed of historical agency and abandoned to the "mute thingness" of nature, the human fallen to the state of the animal (Santner 2006, XV).

But is the creaturely necessarily mute? Or is it rather that creaturely matter speaks in ways we have yet to listen to or understand. Insofar as the creaturely is produced by forces of exposure, might it also bear witness to those forces? In what follows I want to pursue a line of thinking that links a "creaturely memory" to issues of traumatic memory in order to question the ways that bodies hold on to or respond to elements of the past that resist being spoken or written or that the mind is simply unable to retrieve. Memory, of course, has always had a fraught relation to the writing of history because of its unreliability. On the one hand the idea of a creaturely memory comes close to what Pierre Nora has called "true memory"—that which lives in gestures, habits, and unspoken traditions and that he distinguishes from the cultural memory that is stored in archives (Nora 1989, 13). On the other hand, insofar as memory is not only what we can recall but what we are subjected to, we might think of it itself as a kind of wounding, a reminder of the physical vulnerability we share with other creatures, and oftentimes, inflict upon them. For critics like Walter Benjamin, attention to the wounding of other animals lay behind his rejection of humanism, as it also steered him toward what Beatrice Hanssen describes as a "new ethical modality," one that consists of "an extreme attentiveness characterized by a radical openness to the creaturely" (Hanssen 1998, 6). This openness to the creaturely is especially apparent in Benjamin's writings on Kafka, where, moreover, he claims that "animals are receptacles of the forgotten" and adds that "Kafka did not tire from picking up the forgotten from animals" (Benjamin 1969, 132).

What might it mean to pick up the forgotten from animals and how might we do so? To be sure, and as I have written elsewhere, Kafka associates the forgotten with a kind of wounding and also with a superficial manner of healing. That healing is essentially a forgetting of the moment and cause of the wound, a forgetting also of what life was like before the wound. This is the situation of Red Peter in "A Report to an Academy," whose scars are the only traces he has of his former life as an ape, a life he is otherwise unable to remember or speak about (see Weil 2012, 3–22). Red Peter is not that beast whom Nietzsche describes as living "unhistorically" in the present moment. He is rather the humanized beast who has had to forget in order to find "a way out" of his subjection; forgetting is, Nietzsche states, "the property of all action" (Nietzsche 1957, 5–6). The forgotten is thus also endemic to our history. Might an openness to the creaturely help us retrieve what was forgotten and so lost to that history?

It is with these questions in mind that I turn to the work of the Belgian artist Berlinde De Bruyckere and to her 2016 installation in the Hauser & Wirth Gallery in New York, where the viewer is confronted with the affective power of creaturely remains; their power to point toward the very forces that shaped them, historically as well as artistically. There in the wide gallery spaces were a series of life-size horse and hide sculptures, made with wood, wax, iron, fabric, and horse skins. Entitled "No Life Lost," the exhibition was meant to be a tribute to those horses whose lives were lost during the First World War, whether in battle or in transport of men and artillery. The show was also an effort to come to grips with De Bruyckere's own childhood memories of horse slaughter. Her father was a butcher, and while horses in Belgium, as elsewhere in Europe, were often looked upon as noble animals, they were also meat. One area of the gallery thus took on the image and scale of the refrigerator rooms of slaughter houses filled with a series of disembodied skins, here cast in wax, splayed and hung on hooks—a visual reminder of the size and color of each animal who was to be eaten. Having once visited a skin trader's workshop in Anderlecht, Belgium, De Bruyckere remarked, "I was unable to avert my eyes. . . . I knew I had witnessed something I needed to transcribe. Never have I experienced death and life as intensely as I did there."[1] That simultaneity of life and death is what the viewer experiences before the display of waxed skins as mementos of living and breathing horses. They are also remnants of creatureliness, in Santner's terms, the remnants of exposure to systems of power; in this case, those of the food and agricultural industries. As one of De Bruyckere's reviewers wrote, they are "works born of bloodied memory," hers in particular, and ours by implication (Stephens 2012).

History and memory thus join roles in the Hauser & Wirth exhibition as they do within the larger tradition of taxidermy as an art of the trophy and a commemoration of human-animal encounters. But De Bruyckere's is taxidermy of another sort: the arrangement (taxis) of skins (derma) without their bodies as a means to expose rather than participate in anthropocentric practices of trophy hunting. Writing about taxidermy's place within natural history and museums in the nineteenth century, Donna Haraway describes its particular contributions to patriarchal and colonial missions that seek not only to take control of animals from exotic or faraway places but to create visual narratives of that possession that can be displayed for posterity. Taxidermy, she writes, "was made into the servant of the 'real'." It was "the craft of remembering this perfect experience. Realism was a supreme achievement of the art of memory, a rhetorical achievement crucial to the foundations of Western science" (Haraway 1984–1985, 37). This was, in part, because taxidermy came into popularity at the same time as wilderness and the animals within it were receding from view due to increasing urbanization and industrialization. A crucial part of the taxidermist's story, however, was of the killing of the animal itself and which "had to be accomplished as a sportsmanlike act" (Haraway 1984–1985, 36–7; see also Aloi 2012, ch. 2). The animal, in other words, had to be a worthy foe—a perfect specimen of an idealized or perhaps exoticized Nature, conquered and brought back into view only by the superior capacities of man.[2]

More recently, taxidermy has become of renewed interest, not so much for its participation in the subjugation and realist representation of nature but rather for the ways in which its use destabilizes our relations to animals and to art. Rachel Poliquin writes of the irresolvable tension between the live animal taxidermy recalls and the lifeless object it has become. "Animal or object? Animal and Object?" (Poliquin 2012, 5). We are not allowed to choose, she argues. Giovanni Aloi has coined the label "speculative taxidermy" for recent practices that are concerned with the materiality and manipulation of animal skins, but also with "the biopolitical registers inscribed in such operations" (Aloi 2018, 25). Contrasting speculative taxidermy with "botched taxidermy" as theorized by Steve Baker where barely recognizable specimens were represented as passive and speechless victims, Aloi focuses on the productive potential of this new practice in which the skins are "anything but dumb" (Aloi 2018, 25). What they speak about, furthermore, is generated through a host of material/discursive forces that question the boundaries between human and non-human in order to reveal their entangled and interdependent histories.

Crucial for such taxidermied art is that animals are not killed in or for the process. Rather the skins are taken from animals who have died by other causes.[3] Its practitioners are often women, and as with De Bruyckere, the animals represented are often domestic animals, those who were unworthy objects for traditional taxidermy. The very issue of domestication, furthermore, brings to the fore connections between histories of gender, class and species, often rendered in a recognizable stance of submission, if not shame. As we shall see in more detail, such stances establish very different relations between viewer and animal than those seen in taxidermy's traditional celebration of conquest. De Bruyckere pushes to an even further extent the confusion between animal and object that Poliquin articulates, in order to ensure a certain illegibility in the animal it/him/herself and an uncertainty about who or what this animal is or how specifically a horse came to be in this position. The confusion extends even to the viewer's emotional response, which can shift from sympathy or sadness and to disgust.

Horses have been a focus of much of my work over the years, in part, because I wish to recall a time when they were once so central and crucial for our own lives but also because, even now, they continue to raise social, political, and ethical concerns regarding our relations to other animals and to our own animality, or creatureliness. This is especially true of De Bruyckere's horses who display none of the majestic or noble stances traditionally associated with the animal that the eighteenth-century French naturalist Georges Louis Leclerc, Comte de Buffon (1707–1788) described as the "most beautiful conquest of man." One of the especially moving pieces in the installation shows a case containing two life-size horses lying one atop the other as if packed for transport (whether to war or to the butcher) (Image 9.1). Rather than noble conquest, we witness a

Image 9.1 No Life Lost II, 2015. 2015 Horse skin, wood, glass, fabric, leather, blankets, iron, epoxy. 237.5 × 342.9 × 188 cm / 93 1/2 × 135 × 74 inches.

Source: Copyright Berlinde De BruyckereCourtesy the artist and Hauser & Wirth Photo: Mirjam Devriendt.

sad tenderness in the way the bodies touched each other, as if each was protection for the other against the other's creatureliness, its "exposure" to power in Santner's terms.

But the creaturely in De Bruyckere's works also reminds us of what Beatrice Hanssen (1998, 152) saw as Walter Benjamin's efforts to reinscribe the theological into the seemingly abject realm of the creaturely[4] (Image 9.2). Thus, one of the sculptures figures the horse as sacrificial lamb, its legs and life force bound in martyrdom, as the eyes are bandaged as if to prevent it from seeing or acknowledging its exposure. We, the spectators, however, act both as witness and perpetrator insofar as our need for redemption is often sated through animal sacrifice. Seeing this sculpture, I was reminded of the ending of J.M. Coetzee's novel *Disgrace*, where the main protagonist, David Lurie, "gives up" the one dog he has been caring for, the one who he knows, would die for him (Coetzee 1999a, 215). "Bearing him in his arms like a lamb" (Coetzee 1999a, 220), he determines his time has come and offers him up to be euthanized with the others who have been abandoned for being too much or too many. What, I began to wonder, is given or given to be forgotten when we give up the animal.

I did not know at the time that De Bruyckere had been collaborating with Coetzee for some years. At her request, he helped curate her show at the 2013 Venice Biennale, and the two have co-authored a number of books. One, in particular, is important for me here. Its title, *We Are All Flesh*, seems to defy the statement made by another of Coetzee's

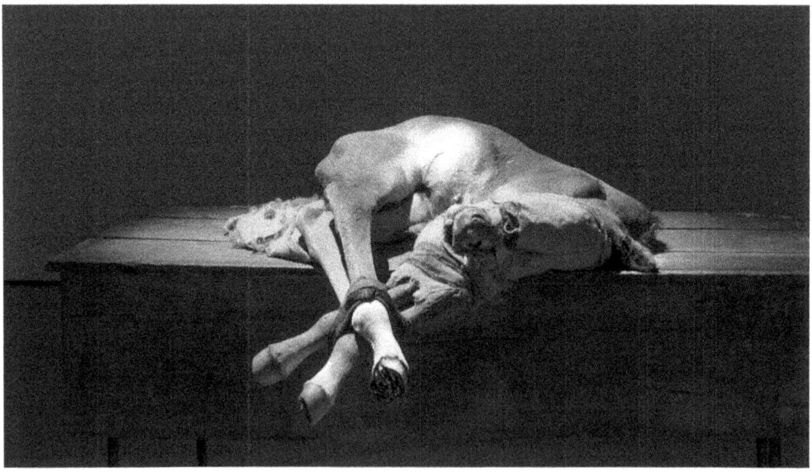

Image 9.2 To Zurburan, 2015. 2015. Horse skin, fabric, wood, iron and epoxy 116.8 × 160 × 127 cm / 46 × 63 × 50 inches.

Source: Copyright Berlinde De BruyckereCourtesy the artist and Hauser & Wirth Photo: Mirjam Devriendt.

Image 9.3 Actaeon, 2011–2012. 2012. Wax, wood, fabric, epoxy, iron armature 142 233.5 80 cm / 55 7/8 × 31 ½ inches.

Source: Copyright Berlinde De BruyckereCourtesy the artist and Hauser & Wirth Photo: Mirjam Devriendt.

protagonists, Elizabeth Costello, who distinguishes human from animal by claiming that the animal's "whole being is in the living flesh" (Coetzee 1999b, 65). *We Are All Flesh*, by contrast, is inclusive of human and non-humans alike and suggests, furthermore, that we humans might have something to learn from the life-worlds in which we are entangled with non-humans. Indeed, this entanglement was at play also in the enormous centerpiece from the Biennale entitled *Cripplewood* (Image 9.3) and which was also on view at Hauser & Wirth. A huge, sculptural composition composed of wax, wool, wood and iron, shaped like intertwined and knotted body parts or possibly tree limbs, its composition demonstrates the shared creaturely and vulnerable exposure of the human and more-than-human worlds.

Memory is a key theme in *We Are All Flesh*. The book is composed of selected passages quoted from Coetzee's many novels—from *Age of Iron* though to *Disgrace*—interspersed with images from De Bruyckere's many installations. As the book jacket explains, the point was not for Coetzee to "write about the objects" but rather to add to them by "juxtapos [ing] the work with creative texts." The resonance between the visual images and the writing is powerful not because of any descriptive or explanatory function but rather because of the unexpected questions one raises to the other. The first two images in the book were taken from De Bruyckere's "The Wound" of 2012 (Image 9.4) and "Lingam II" of 2013 and focus on what look like scars created and exacerbated by the stitching over of wounds, whose flesh lies lumpy and reddened beneath. The first few quotations from Coetzee also reference the "scars of suffering" which "we" (referring probably to we humans although I'm not really sure) carry with us. With these scars, he writes one "develop[s] a hard skin against one's honesty" (De Bruyckere and Coetzee 2013, 5). The fourth excerpt is taken from *Disgrace*. It references the rape that Lucy, daughter of the

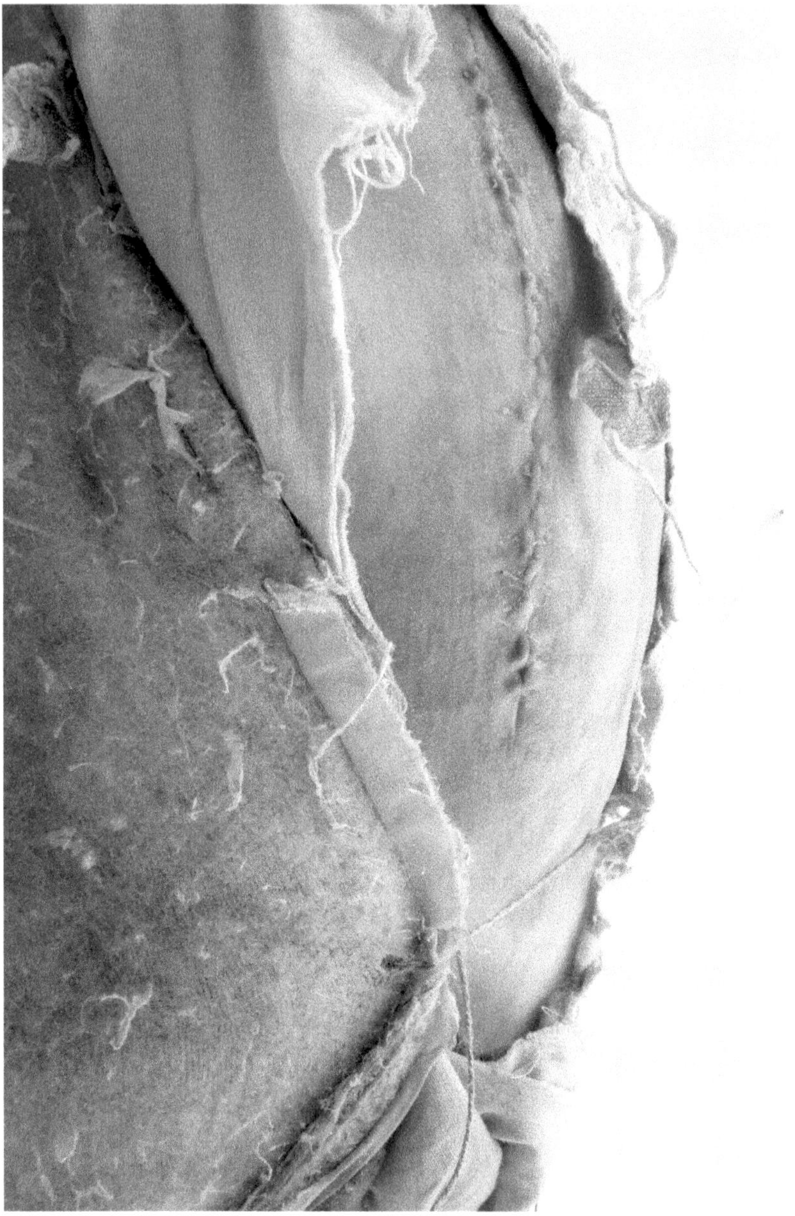

Image 9.4 The Wound I. 2011–12 (detail) 2012. Wax, epoxy, iron armature, leather, wood, horsehair, fabric, blanket. 125 × 116 × 41 cm / 49 1/4 × 45 5/8 × 16 1/8 inches.

Source: Copyright Berlinde De BruyckereCourtesy the artist and Hauser & Wirth Photo: Mirjam Devriendt.

main protagonist, suffers in her own home, while her father lies helpless and beaten in a bathroom nearby.

> So time does indeed heal all. Presumably Lucy is healing too, or if not healing then forgetting, growing scar tissue around the memory of that day, sheathing it, sealing it off. So that one day she may be able to say, "The day we were robbed," and think of it merely as the day when they were robbed.
>
> (6)

Healing, here, implies a forgetting. It is a stitching or writing over of past events effected in order to remember only those parts of the past with which one can more comfortably live. Memory thus becomes history by allowing essential, unknowable moments to slip to the wayside, or to unconscious oblivion, suggesting that history too, perhaps like Nietzsche said of life, is impossible without forgetting (see Nietzsche 1957, 5–7.)

Whereas *We Are All Flesh* opens with images of human scarring, towards the very end of the book we find, somewhat surprisingly, images of the glossy, taxidermied coat of a horse, shot from angles that render the equine's curled or folded body at once glossy and tortured (Image 9.5). On the one hand, we might wonder if this luminous fur is another kind

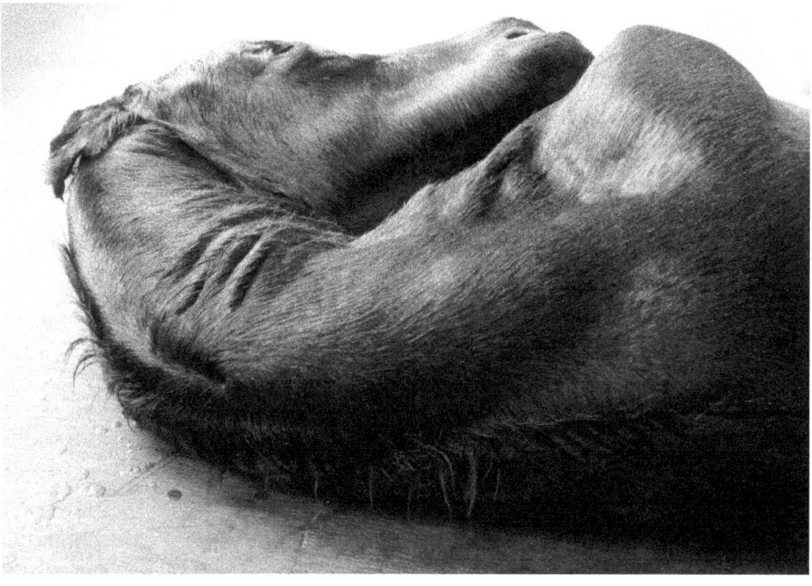

Image 9.5 Lost II (detail) 2007. Horseskin, Apoxy, Metal Wood. 98 × 151.5 × 164 cm / 38 5/8 × 59 5/8 × 64 5/8 inches.

Source: Copyright Berlinde De BruyckereCourtesy the artist and Hauser & Wirth Photo: Mirjam Devriendt.

of writing, one meant to distract us from the invisible wounds involved in taxidermy's history. What else, then, might be the relation of this fur to the matter of scarring? Do horses not scar and heal in the same way? Might this be one reason that Walter Benjamin describes animals as "receptacles of the forgotten," suggesting thereby that animals are unable to write over the past—as Lucy did—which instead remains stored within their bodies? If so, can those forgotten elements be retrieved or "picked up," as Kafka did according to Benjamin? Indeed, it was Kafka's attentiveness towards all living creatures, which Benjamin, citing Malebranche, praised as the "natural prayer of the soul" (Benjamin 1969, 134).

The work of Nicolas Malebranche, a late-seventeenth-century theological philosopher, may have particular relevance for our thinking about creaturely memory. Malebranche broke from the Cartesian tradition to suggest that an ideal intelligibility can be had through sentient experience and not, as Descartes would have it, through the mind or language (or at least of language as we conceive of it). In an essay on Malebranche and Merleau-Ponty, Judith Butler writes that what one finds in the work of both philosophers is an insistence on "the centrality of the body to the act of knowing" (Butler 2015, 39). The knowledge disclosed through the body, however, is difficult to bring to light, Butler explains, because of the temporal delay between sensing and knowing and because of the incommensurability between sensing and speech. "Bodily significations do not become successfully converted or sublimated into speech [. . .] even as the two modalities remain related to one another, even if in symptomatic ways" (Butler 2015, 15). In a similar vein, De Bruyckere's sculptures and Coetzee's writings are symptomatically related, but their symptoms are not always translatable or, even, intelligible. Coetzee writes about de Bruyckere saying, "I have been touched by her work in ways that are often obscure to me. I would not wish it otherwise."[5] Touch, Coetzee thus implies, speaks but is not always intelligible.

The point here is that a knowing, animal body is often incommensurate with the I who speaks. Certain bodily and traumatic experiences, like that which Lucy undergoes in *Disgrace*, may never be fully recoverable in speaking or writing, and this may be because our language and writing are always somewhat "out of touch," with the body. According to Butler, touch for Malebranche and Merleau-Ponty is the "animating principle of feeling and of knowing" (Butler 2015, 42). It is prior to but also the very condition for my saying "I," or at least that is the case for any "I" that feels and has a bodily experience. Insofar as that feeling begins with a touch that is not mine, it can also threaten me in the very boundaries I imagine for myself. Indeed, it is not uncommon to "lose trust in touch," as does the young protagonist in Mary Gaitskill's novel *The Mare*. The aptly named Velvet begins to shield herself from physical contact with the humans who have abused her, physically and verbally because "she had lost her trust in touch" (Gaitskill 2015, 315). Part of the healing process for Velvet is to find herself drawn to a bodily contact with horses, those creatures, as the trainer and philosopher Vicki Hearne describes, whose "intelligence we can just barely have any understanding of, partly because we communicate with horses through touch. [. . .] Every muscle twitch of the rider will be like a loud symphony to the horse" (Hearne 2000, 107–8). The point is to make oneself legible to the horse and so to come to an understanding that can only happen through the skin, and not through words. Horses speak by other means, if we are willing and able to listen and to hear.

The incommensurability of language and touch has important implications not only for human-animal relations but also for human-human relations. Considering the nature of sensation, Butler asks under what conditions does the "I become capable of reporting on what it feels?" (Butler 2015, 42). Can we report on touch outside of the conditions that our language and conventions impose? These questions have ethical implications when we ask about the ways in which we may begin to read the imprint and reception of touch on and by another—human or animal. How and where does the animal/body speak? In asking this question, I am reminded of the efforts of French feminists of the 1970s to write the body and to do so in a language that could somehow be outside of patriarchal impositions. "Je suis là ou ça parle," "I am there where it speaks," Hélène Cixous wrote, in the hopes of merging a written I with the speaking and remembering body referred to as "ça" or "it"—which speaks otherwise than in conventional language (Cixous 1977, 488).[6]

This "ça," which might refer to a bodily "id," has become the focus of recent writings about trauma and healing, specifically about the insufficiency of talk therapy to fully heal the body, even as it may offer the balm of a better story about the past. Dr. Bessel Van der Kolk writes of that extraordinary capacity of the mind to rewrite memory, but he also warns that the rewriting may not reach the actual, deeper memory which lives in the body, such as we saw with Lucy's scars referenced earlier. "Perhaps the most important finding in our study was that remembering the trauma with all its associated affects, does not, as Breuer and Freud claimed back in 1893, necessarily resolve it. Our research did not support the idea that language can substitute for action," writes Van der Kolk, in his 2014 book, *The Body Keeps Score: Brain, Mind and Body in the Healing of Trauma* (196). Rather than Freud and Breuer, Van der Kolk looks back to Pierre Janet, Charcot's co-researcher at the Salpêtrière in Paris. It was Janet who pointed to the difference between "narrative memory"—the stories people tell about trauma—and traumatic memory itself (Van der Kolk 2015, 181). Remembering, he asserts, cannot resolve trauma, because "the imprints of traumatic memories are not organized as coherent logical narratives but in fragmented sensory and emotional traces: images, sounds, and physical sensations" (Van der Kolk 2015, 178). Traumatic memories, that is to say, are stored in the body, expressed physically, and so it is the body that must also be healed.

"Perhaps it is so that all languages are, finally, foreign languages, alien to our animal being," writes Coetzee (De Bruyckere and Coetzee 2013, 49). In what might be an illustration of this point he—or his narrator—comments upon how we speak of

> *the dog with the sore foot or the bird with the broken wing*, but the dog does not think of itself in those terms. Or the bird. To the dog when it tries to walk, there is simply *I am pain*, to the bird when it launches itself into flight, simply *I cannot.*
>
> (Ibid., 50)

Animals are their body (or their flesh), which does not imply that they have no consciousness or subjectivity but rather affirms that the relation between subjectivity and bodily being for them is not one of separation and ownership as of property, but a kind of cathexis. Coetzee writes that language "cannot get purchase, cannot get going until it has split up the unity of experience" (*Ibid.*, 15). Healing, then, must work by re-establishing the unity of body and consciousness that our language, our autobiographical stories have split (Van der Kolk 2015, 238).

In this regard, we might see De Bruyckere's horses saying, "I am flesh," "I am/we are pain," "I am shame," or "I am fur," as taxidermied hides are presented in various poses of slaughter or sacrifice, with faces bandaged, legs tied, and bodies generally and variously incapacitated, if not skinned in preparation for eating. Their presence and present tense express an embodied past and continuity with the past trauma. Coetzee has written elsewhere that her sculptures "explore life and death, death in life, life in death, life before life, death before death—in the most intimate and disturbing way. They bring illumination, but the illumination is as dark as it is profound."[7] Such dark illumination is intimately related to the contradictory power of the fur—fur that was part of the attraction for the artist when the horses were still alive and was taken only after their death. Darkness and illumination, death and life gleam in this fur that is rendered both by industry and by theology to warn of the aesthetic exploitation of animal bodies.

In his book on *Surface Encounters*, Ron Broglio also asks "how we might think from the fur of the beast" and whether such a surface thinking might work to invert the hierarchies of surface and depth with which we have judged animal and human worlds (Broglio 2011, xv). Might a focus on the fur help us to think beyond our humanistic frames of reference, beyond our prioritization of the metaphysical? It is curious how many of the images in *We Are All Flesh* show us fur separated from the bodies that it is expected merely to cover or belongs to—as if for de Bruyckere, the fur were a world of its own. Fur, moreover, is contrasted with the many images of Cripplewood, whose various parts or limbs, whether covered in fleshy fabric or flayed as skinless bones, are at once abject—bloodied or broken or torn—and threatening—sharp, pointy, and

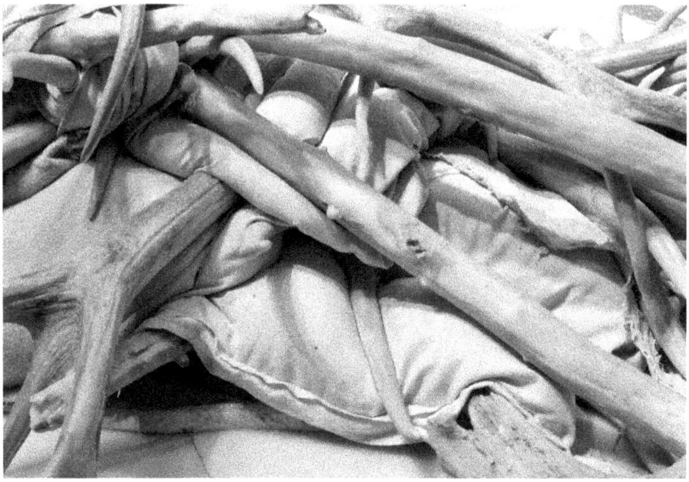

Image 9.6 Actaeon II (Hamam), 2012 (detail) 2012. Wood, fabric, apoxy, iron (dimensions variable)

Source: Copyright Berlinde De BruyckereCourtesy the artist and Hauser & Wirth Photo: Mirjam Devriendt.

piercing (Image 9.6). The fur is totally other. Soft and glossy with swirls of luminous depth, it invites touch, tempts us to get lost in its feel, "Lost" being the title of a separate 2004 installation from whose sculptures the shots of fur were taken. Thus, the image of K36 (The Black Horse) shows a close-up of a glossy, gleaming coat on a body whose form we cannot quite read—perhaps a horse in the process of lying down or standing up. Reattached to their sculptural bodies, the fur becomes again the surface of a body that reads now as distorted, bent over, or hunched, as if referencing the hunchback that, according to Benjamin, will disappear with the coming of the Messiah (Benjamin 1968, 134)[8] (Image 9.7). This is a figure of shame and, as such, illustrates in a way that the fur alone does not, the juxtaposed passage from Coetzee's *Age of Iron*.

> Perhaps I should simply accept that that is how one must accept to live from now on: in a state of shame. Perhaps shame is nothing more than the name for the way I feel all the time. The name for the way in which people live who would prefer to be dead.
>
> (78)

Coetzee's female, dying narrator thus joins De Bruyckere's horse in an experience of the creaturely, not as a fallen state peculiar to humans, as Santner would have it, but rather as one shared by humans and non-humans alike. The pose of this horse recalls that of William Wegman's dog

Image 9.7 K36 (The Black Horse). 2003 Polyurethane foam, horse hide, wood, iron, 295 × 286 × 158 cm / 116 1/8 × 112 5/8 × 62 1/4 inches.

Source: Copyright Berlinde De BruyckereImage Courtesy Saatchi Gallery, London.

in his photo, entitled "Bikini." Reading this image as a figure of shame, Alice Kuzniar explains that the word comes from the Indo-European word *skam or skem*, meaning hide or skin (Kuzniar 2006, 86). Shame is the result of the manner in which we turn fur to flesh by dressing it in clothes and framing the body such that what could and should be nude (in the terms of classical aesthetics) becomes naked and wanting. But shame is also, Kuzniar reminds us, an interpersonal, if not the most interpersonal, of affects (Kuzniar 2006, 71). In a tradition reaching back to Sartre and before, shame is dependent upon the regard of the other: I see myself as the other sees me. Shame is thus another manifestation of the ocularcentrism of Western philosophy and of human relations more generally. Perhaps this is one reason Darwin cites shame as the most human of emotions, one which animals show little signs of. Animals show no need to conceal their "sex" nor their excrement, and this, according to Freud, is the reason the dog has earned contempt among humans and become a term of abuse (Freud 1978, 99–100).

Might the images of fur here exposed, not as covering for a body, not as separable from it, but as an essential component of the creaturely, also offer a different pathway toward making the body intelligible? This way would depend not on vision, not on form, but on feel and touch—if not also on smell—a sense of which humans are notoriously impoverished.[9] If the dog or horse "palpably embodies its creatureliness," the way to that creature must be through the body and by touch (Kuzniar 2006, 71). And touch may also be the means to get to a creaturely memory that precedes or has not been clothed in writing.

In an essay on the "Fecundity of the Caress," Luce Irigaray asks: "Is the memory of touching always disguised by senses that forget where they come from? Creating distance through a mastery that constitutes the object as a monument built in place of the subject's disappearance?" (Irigaray 1993, 215). Mastery through distance is an all-too-common way of reacting to disgust and to the fear of vulnerability that, as Walter Benjamin suggests in a short reflection "On Gloves," is effected through touch. "All disgust begins with disgust in touching"—touching that risks revealing our kinship with animals, a kinship, a contact, a touch that impels man to "make himself its master" (Benjamin 1978, 66–7). Such mastery, Coetzee suggests in a passage from his own autobiographical work, *Boyhood*, however, may be a refusal of life.

> Nothing can touch you, there is nothing you are not capable of. Those are the two things about him, two things that are really one thing, the thing that is right about him and the thing that is wrong about. Him at the same time. This thing that is two things means that he will not die, no matter what; but does it not also mean that he will not live?
>
> (De Bruyckere and Coetzee 2013, 51)

To distance oneself from touch is thus to avoid recognition of our vulnerability, of our essential relatedness to others—human and non-human. It is to deny our creatureliness. And this may be why the memory of touching, the memory of the flesh is, Irigaray insists

> the most insistent and the most difficult to enter into memory [. . .] Memory of the flesh, where that which has not yet been written is

> inscribed, laid down? That which has no discourse to wrap itself in? That which has not yet been born into language? That which has a place, has taken place, but no language. The felt, which expresses itself for the first time. Declares itself to the other in silence.
> (Irigaray 1993, 215)

Like the flesh, but without its associations with nakedness or shame, the fur is also a site of what has place or has taken place but has not yet been inscribed in language. De Bruyckere's fur expresses an invitation to touch and thereby to remember or to reach a past event that might otherwise be forgotten or written over. The point here is not that animals don't know shame, nor that it is only through a Nietzschean animal forgetting that the creature heals—whether from shame or from other traumas. Knowing not "the meaning of yesterday or today," animals, in Nietzsche's world, would not have the memory to know or embody shame (Nietzsche 1957, 5). If, however, we understand a creaturely memory to be stored and exposed otherwise in and on the body, then it may be through such contact zones of flesh and fur that we can begin to find alternative methods for working through shame and the experience of vulnerability that gives rise to it.

This is not a new idea, of course, and is one reason why animal therapy programs have often been successful, programs which, however, only rarely consider their effects on the animals. In my recent work on horses in nineteenth-century France, I examine how theories of training horses figured as a means for reworking bodily responses to memories—whether equine or human.[10] If the body keeps score, as Van der Kolk says, it is the body which needs to find new ways of responding to stimuli and situations that, in the past, left it vulnerable. Touch is that place where being acted on and acting are simultaneous, the place where vulnerability can turn to agency if also vice versa. And yet, as Hearne insists, touch is a kind of "intelligence" of which we have little understanding (Hearne 2000, 106–7). This is why we must learn to read a horse's skin and become, in turn, "kinesthetically legible" to the horse. We depend upon the visual for intelligibility, but horses cannot be read visually, "without knowing them kinesthetically" (ibid., 110). We must learn to read the fur, learn how to make the tactile or the haptic intelligible. This is one aim of a "haptic visuality," whose urgency takes us from embodied, visual criticism to creaturely relations.[11]

To the extent that De Bruyckere, and taxidermy more generally, plays with reactions to flesh and fur, it reminds us that one does not fully live in the world or "have a world" without touch. Signs in the gallery saying "please do not touch the artworks" make our yearning for a tactile experience all the more apparent in its interdiction. "If," Butler writes, "I can come to touch and feel and sense the world, it is only because this 'I' before it could be called an 'I,' was handled and sensed" (Butler 2015,

11). To be sure, touch makes one vulnerable to the adverse recognition of the other. I may be drawn to stroke or scratch the fur of horses, dogs, cats, knowing that my touch may provoke a bite or kick or scratch. But it might also be a first step toward a new reaction, a new ground for two creatures to become together. How can we re-establish trust in touch? This seems to me to be the question that Coetzee and De Bruyckere raise in their very different, but mutually inflective, languages. To be sure, certain words can also touch and move affectively, as Elizabeth Costello herself emphasizes in describing herself as "an animal exhibiting, yet not exhibiting, to a gathering of scholars, a wound which I cover up under my clothes but touch on in every word I speak" (Coetzee 1999b, 26). To expose the wound is to be vulnerable, but it is also to make her and others aware of her animality, her creaturely being. The point is not to forget, excise, or stitch over the past, not to dissociate the self, nor to master the other, but, rather, to accept our shared animal fragility as a meeting place for new beginnings and becomings. It is to give ourselves over to the worlds enveloped by the fur, in order to relearn how to feel and so turn disgust into wonder, not mastery. No small order, I know (Image 9.8).

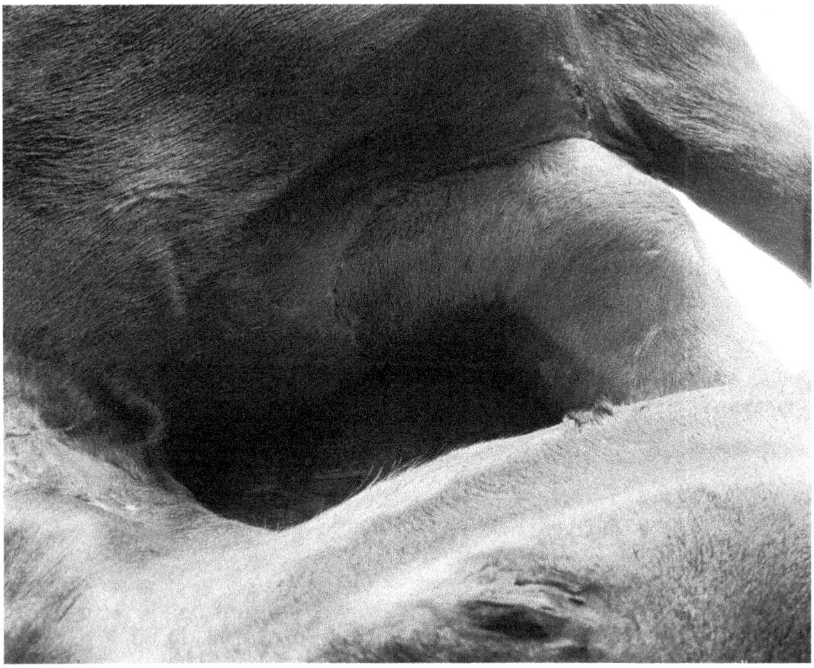

Image 9.8 Lost III, 2011 (detail) 2011. Horse skin, textile, wood, epoxy, iron. 125 × 160 × 165 cm / 49 1/4 × 63 × 65 inches

Source: Copyright Berlinde De BruyckereCourtesy the artist and Hauser & Wirth Photo: Mirjam Devriendt.

Notes

1. Berlinde De Bruyckere, *No Life Lost*. www.hauserwirth.com/exhibitions/2719/berlinde-de-bruyckere-no-life-lost/view/ (last accessed: 4/21/18).
2. Taxidermied animals, writes Aloi, acted as referent for "the exotic and marvelous life forms that inhabited faraway lands," adding that for the Victorians, these bodies thus "functioned as place holders in the mapping of the land incorporated by the Empire (Aloi 2012, 27).
3. De Bruyckere is careful only to use skins of horses salvaged from a veterinary school so that there would be no bloodshed in the name of art. See Olda (2013).
4. Hanssen, *Benjamin's Other History*, 152.
5. Berlinde De Bruyckere, "Kreupelhout-Cripplewood," www.belgianpavilion.be/en/2013/artists/berlinde-de-bruyckere (last accessed 4/20/18).
6. Hélène Cixous, "Entretien avec Françoise van Rossum-Guyon," *RSH* 168 (December 1977).
7. Cited in de Bruyckere, "Kreupelhout."
8. As Santner explains, Benjamin writes of Kafka's works as being haunted by the figure of the hunchback, who "bows his head far down on his chest" from the weight of the labor and punishments imposed upon him. "The figure is crucial for Benjamin for he also suggests that redemption can be understood as a passage through and beyond the creaturely life materialized in these cringed bodies" (Santner 2006, 25).
9. As Virginia Woolf put it in *Flush*, "The human nose is practically non-existent. The greatest poets in the world have smelt nothing but rose on the one hand, and dung on the other. The infinite gradations that lie in between have gone unrecorded" (Woolf 2009, 86). Vicki Hearne also coins the word "scent-blind" for humans since we don't have a "sense of scent" and are useless at tracking (2000, 80).
10. It was especially during the nineteenth century that memory, like heredity, was understood to be inscribed in a body and, consequently, that the body needed to be trained to resist inherited impulses. (See Weil, 2020)
11. I borrow the term "haptic visuality" from Marks (2000, xi).

References

Aloi, Giovanni (2012). *Art and Animals*. London: I.B. Tauris.

Aloi, Giovanni (2018). *Speculative Taxidermy: Natural History, Animal Surfaces and Art in the Anthropocene*. New York: Columbia University Press.

Benjamin, Walter (1969). "Franz Kafka." In *Illuminations*, 111–45. New York: Schocken Books.

Benjamin, Walter (1978). "One Way Street." In *Reflections*, 61–94. New York/London: Harcourt, Brace, Jovanovich.

Broglio, Ron (2011). *Surface Encounters: Thinking with Animals and Art*. Minneapolis: University of Minnesota Press.

Butler, Judith (2015). *Senses of the Subject*. New York: Fordham University Press.

Cixous, Hélène (1977). "Entretien avec Françoise van Rossum-Guyon." *RSH* 168 (October–December): 479–93.

Coetzee, J.M. (1999a). *Disgrace*. New York: Penguin.

Coetzee, J.M. (1999b). *The Lives of Animals*. Princeton: Princeton University Press.

De Bruyckere, Berlinde (2013). "Kreupelhout-Cripplewood." Accessed 20 April 2018. www.belgianpavilion.be/en/2013/artists/berlinde-de-bruyckere.

De Bruyckere, Berlinde, and Coetzee, J.M. (2013). *We Are All Flesh*. Ghent: MER, Paper Kunsthalle.

Freud, Sigmund (1978). "Civilization and Its Discontents." In *The Standard Edition of the Complete Psychological Works of Sigmund Freud*, 57–145. London: Hogarth Press.

Gaitskill, Mary (2015). *The Mare*. New York: Pantheon.

Hanssen, Beatrice (1998). *Walter Benjamin's Other History: Of Stones, Animals, Human Beings and Angels*. Berkeley: University of California Press.

Haraway, Donna (1984). "Teddy Bear Patriarchy: Taxidermy in the Garden of Eden." *Social Text*, no. 11 (Winter): 20–64.

Hearne, Vicki (2000). *Adam's Task: Calling Animals by Name*. New York: Common Reader Edition/Akedine Press.

Irigaray, Luce (1993). "The Fecundity of the Caress." In *An Ethics of Sexual Difference*, translated by Carolyn Burke and Gillian C. Gill. Ithaca: Cornell University Press.

Kolk, Bessel Van der (2015). *The Body Keeps Score: Brain, Mind and Body in the Healing of Trauma*. New York: Penguin Books.

Kuzniar, Alice (2006). *Melancholia's Dog: Reflections on Our Animal Kinship*. Chicago: University of Chicago Press.

Marks, Laura (2000). *The Skin of Film: Intercultural Cinema, Embodiment, and the Senses*. Durham: Duke University Press.

Nietzsche, Friedrich (1957). *The Use and Abuse of History*. Translated by Adrian Collins. New York: Macmillan Publishing.

Nora, Pierre (1989). "Between Memory and History: Les Lieux de Mémoire." *Representations*, no. 26 (Spring): 7–24.

Olda, Danny (2013). "The Nightmarish Horses of Berlinde De Bruyckere." *Beautiful Decay*. Accessed April 20, 2018. www.beautifuldecay.com/2013/03/13/the-nightmarish-horses-of-berlinde-de-bruyckere/.

Pick, Anat (2011). *Creaturely Poetics: Animality and Vulnerability in Literature and Film*. New York: Columbia University Press.

Poliquin, Rachel (2012). *The Breathless Zoo: Taxidermy and the Cultures of Longing*. University Park: Penn State University Press.

Santner, Eric (2006). *On Creaturely Life: Rilke, Benjamin, Sebald*. Chicago: University of Chicago Press.

Stephens, Andrew (2012). "Works Born of Bloodied Memory." *The Sydney Morning Herald*, June 2. www.smh.com.au/entertainment/art-and-design/works-born-of-bloodied-memory-20120601-1zmd6.html.

Weil, Kari (2020). *Precarious Partners: Horses and Their Humans in Nineteenth-Century France*. Chicago: University of Chicago Press.

Weil, Kari (2012). *Thinking Animals: Why Animal Studies Now?* New York: Columbia University Press.

Woolf, Virginia (2009). *Flush*. London: Oxford World Classics.

10 Animals as Monsters, Victims, and Models in the Age of Decadence[1]

Eva Voldřichová Beránková

1. Introduction

At first glance, it may seem paradoxical to wonder what role animals played in French Decadence. This late-nineteenth-century artistic and literary movement followed an aesthetic ideology of excess and artificiality. In schoolbooks, it is often characterized by self-disgust, sickness of the world, general scepticism, delight in perversion, the employment of crude humour, and a belief in the superiority of human creativity over logic and the natural world. Inspired by Théophile Gautier and especially by Charles Baudelaire, Decadence in France spread after the defeat of Sedan (1870), when the French elites likened their own country to late Roman antiquity, and their Prussian adversaries to Barbarians. Convinced of the imminent end of France, and even of Western civilization, the Decadents developed a real "aesthetic of the setting sun,"[2] which valued the beauty of the things in decline—a civilization or a landscape, or even a woman, is never as beautiful as when approaching an end, which offers a delicate charm.

The Decadents liked to identify with the aristocratic elite and displayed a bitter contempt for their immediate literary predecessors, the Naturalists, even though some of them (especially Huysmans) started their literary career precisely in Naturalist circles organized by Émile Zola. After 1884, a whole generation of young men rebelled against the positivism, scientific spirit, and materialism of their time, reproaching Naturalists for reducing man to physiology and for ignoring the metaphysical dimension of the world. Their symbolic spokesperson was Jean des Esseintes, the Decadent protagonist of the Joris-Karl Huysmans' novel *À rebours* (quite significantly translated into English as *Against Nature*).

Surprised by the violence of these intergenerational attacks (Zola had always considered Huysmans as his best disciple), the Naturalists struck back and emphasized the morbid nature of the Decadents' creation, as well as the young authors' lack of effectiveness. In the second half of the 1880s, the quarrel was at its height, intensifying and becoming more and more virulent. It was precisely at this moment that animals are employed, since in numerous (not only Naturalist) texts written during the *fin de siècle* period, the Decadents were described as a strange animal species. Paradoxically, they accepted this originally pejorative label, playing with it and using it to their advantage. The Decadents used metaphors and animal comparisons to define themselves on both human and literary levels: exotic monsters and hybrids, monkeys, dogs, molluscs, invertebrates or even insects, and so on; the scale of "strange beings" to which late-nineteenth-century writers identified themselves was very broad and their texts became more and more provocative. This need to constantly confront the animal otherness to find one's own identity is probably one of the most fascinating legacies of the Decadence. In particular, the omnipresent identification with strange animals seems to have allowed the Decadents to introduce a new poetics, to promote a new lucidity regarding the creative process, and to anticipate in a certain way the contemporary gnoseological approaches that question the "anthropological machine" (Agamben 2004). This chapter explores the ways in which the Decadents transcended the human-animal dichotomy, as well as their possible role in challenging the traditional representations of human superiority. It also investigates the strategies they used to avoid the anthropocentric paradigm and the insights into animals' lives and minds that we can deduce from their predominant metaphors, refined and uncanny rhetorical figures and literary devices.

2. When an Insult Becomes a Sign of Identification

In one of his letters to Émile Zola from 1886, Paul Alexis describes the Decadents as "animals making too much noise" (Alexis 1971, 164). Even given the virulence of the polemic between the Naturalists and the Decadents, this was the first time that the former went so far as to deny the latter's humanity. Several months later, Léo Trézenik, one of the offended parties, adopted this definition ironically in his erotic novel *The Skirt*, explaining that in a Parisian pub called Le Hanovre, there is always a Decadent "being displayed at liberty" like an exotic monster. The gentlemen passing by ask with curiosity: "What kind of two-legged-creature is this?," while a tamer warns the crowd about the strict "prohibition against throwing anything to the animal" exhibited on stage (Trézenik 1887, 118–19).[3]

As *The Skirt* is an autofictional novel recounting, not without provocation, Trézenik's life—from the time of his studies in the Breton city of Vannes to his fictional sexual adventures with a bourgeois woman (who may be the mother of the French writer Octave Mirbeau)—the episode of the Decadent is to be understood in the jocular spirit of the Zutists, the Hirsutes, the Hydropaths, or the Jemenfoutists[4] the author was part of. While the Decadents are identified with animals exhibited on the podium, the author himself adopts the pen name of Trézenik, which in Breton means "a musical instrument close to the harpsichord," "a conifer of the Picea genus," "a type of wine," or "a wooden cage used for the fattening of capons." Thus, a "chicken cage" is writing a novel in which humans are treated like circus animals. This *fin de siècle* humour seems rather non-anthropocentric.

In the early 1890s, the insults addressed to the Decadent writers became even more aggressive and specific. In his thesis novel called *Termite*, Joseph-Henri Rosny spoke of the "monstrous monkeys of Decadence" (Rosny 1890, 54). In 1897 an odd treatise appeared, published under the title *Decadent Poetry Through the Eyes of Psychiatric Science*, in which a doctor Laurent described the physiognomy of contemporary writers, conjuring up a morbid museum of fantastic creatures:

> Here is one without a forehead, another suffers from advanced prognatism, three others are affected by an apparent asymmetry of the face. . . . Do you see these plagiocephalic, oxycephalic, acrocephalic heads, these shapeless or twisted noses, these bare and asymmetrical cheeks, these broad ears like badly hemmed handles, these huge zygomas, these heavy and prominent jaws[?]
>
> (Laurent 1897, 121)

In addition to primates, the Decadents were often compared to dogs, with special reference to a biblical verse: "*Sicut canis, qui revertitur ad vomitum suum, sic imprudens qui iterat stultitiam suam*" ("As a dog returns to its vomit, so fools repeat their folly," Proverbs XXVI, 11), as they seemingly devour previous texts without masticating, vomiting out their passages in their own texts (Palacio 2011, 56). This strange metaphor of a "precipitous intertextuality" appeared in many contemporary reflections, including those of the Decadent authors themselves (Émile Bergerat, Léon Cladel, or Joris-Karl Huysmans). It seems there is no shame for a *fin de siècle* writer in being compared to a dog vomiting out its own texts or playing around with others' texts.[5]

Some writers descended even lower down the chain of beings, presenting the Decadents as molluscs. For example, Henri Pivert, the protagonist of Fernand Clerget's eponymous novel, is described as "a genuine Decadent, retracting his brave tentacles," while in the following paragraph, the hero is directly called a "snail" (Clerget 1891, 101 and 161). This metaphor is meant to express both the Decadents' prominent curiosity and their detachment from reality. With their tentacles, they carefully examine a world that is not always made to suit their tastes. At the slightest clash with the brutality of the surrounding environment, the Decadents retreat into their ivory towers as into a shell.

The idea of a progressive decline implied in the Decadent aesthetics often led to a sort of evolution or ontogenesis turned upside down, meaning a regression from primates to canines, from mammals to molluscs, from frogs to tadpoles. As Albert Millaud said:

> No Decadent . . . is the final one. Below him there is always a tadpole that starts to show off and to call himself "held down". This is the beginning of the very last series that goes from the "leached out" via the "liquefied" to the "decomposed".
>
> (Millaud 1887, 178)

This brings us to the last set of metaphors including the invertebrates and insects. In *Les Déliquescences d'Adoré Floupette*, a parody of Verlaine's, Mallarmé's, and Moréas's poetry, the Decadents are turned into tapeworms—in French, *vers solitaires*, for example, "solitary worms" (Vicaire and Beauclair 1923, 65)—while in Gustave Ghèz's story *A Question of The Heart* (*Question du cœur*), their poetics are ridiculed as an amateur "art of tapeworm killing" (Ghèz 1888, 180).

These metaphors are based, on the one hand, on the abundant thematic representations of cemetery scenes and worms in contemporary poetry—for example, on the principle of *memento mori* that the Decadent movement borrowed from the Baroque and developed into entire morbid epics—and, on the other hand, on the allegedly "larval" aspect of the Decadents themselves (Goncourt 1864, 1097–8).

Whether by reference to monkeys, dogs, molluscs, or helminths, comparisons to animals were invariably denigrating and always implying a deficiency. The Decadents were "never sufficiently" rational, effective, healthy, manly, sober, disciplined, beautiful, or courageous to impress their (Naturalist or Parnassian) elders. Even worse, they were always found lacking a little something, unable to become real human beings.

This summary of the image of Decadent writers in contemporary critical texts allows us to shift the focus to the animals with which the authors of the *fin de siècle*, on the contrary, liked to identify. Their list, surprisingly, only partially varies from the caricatures mentioned earlier.

3. In Praise of Hybridity

Obviously, even Decadent texts contained the "usual" poetic fauna such as sensual cats (Baudelaire, Cros, Mallarmé, Rollinat, Verlaine), suffering horses or dogs (Corbière, Laforgue), ethereal swans (Baudelaire, Laforgue, Mallarmé, Rodenbach), mysterious owls (Jarry), bloody tigers (Cros), solitary bears (Laforgue), or vulnerable does (Remy de Gourmont).

Nevertheless, the *fin de siècle* period added other, specifically "Decadent," creatures to this (very incomplete) enumeration of more traditional poetic animals. First of all, there were amphibious or hybrid animals. Contemporary female writers identified themselves with some half-birds—mysterious beings anticipating the future "flight of liberated women." In her educational novel called *One Towards the Other*, the suffragette Louise-Marie Compain described wives leaving their traditional homes for work as an "archeopteryx" (Compain 1903, 251). From these temporary beings, a new Eve should gradually evolve, able to fly freely with her own wings, regardless of men's mocking glares or the humiliating restrictions set by social conventions. At the same time, a new Adam must also appear. He shall be an understanding and courageous being who will accept such a disturbing change. The novel *One Towards the Other* described (in a rather utopian way) the progressive advent of such a man of the future. Indeed, the conflicting relations between Laure Prevel and Henri Debordy had to undergo many dramatic vicissitudes before the young man recognizes his wife as "equal in dignity" and approaches life "like a new Adam advancing on a rejuvenated earth" (Compain 1903, 304). Like a mythical bird, the archaeopteryx symbolically opens the door to the paradise of a sexually reconciled, just, equal, and happy new humanity.

A similar hybrid metaphor was used by Thérèse Bentzon, a highly popular author of the *Belle époque*, whose proses, mostly dedicated to marriages of convenience and subsequent adulteries, attracted readers with a curious mixture of eroticism and conservative morality. In her soliloquies, one of the heroines called Odette hesitates between adultery and women's solidarity: "poor little bat, a hybrid and suspicious animal, your place is neither among birds nor among rats; in the midst of the former, you feel a hidden affinity to the latter" (Bentzon 1889, 132).

Neither female nor male, bird nor rat, feminist writers of the late nineteenth century used animal metaphors to express their difference, their inadequacy in the traditional roles of women, hoping for a profound social metamorphosis which would finally mark their acceptance as heterogeneous, ambiguous, peculiar beings. With such metaphors as the archaeopteryx or the bat, female writers invited men to join them in the adventure of hybridity.

Male writers did not necessarily connect hybrid animals to the criticism of social conventions but rather used the intimate sensation of ambiguity, or even the monstrosity of some parts of their own bodies. In Baudelaire's *The Flowers of Evil*, the poet-albatross was nobly flying in the sky, but after his capture by the sailors:

> This voyager, how comical and weak!
> Once handsome, how unseemly and inept!
> One sailor pokes a pipe into his beak,
> Another mocks the flier's hobbled step.
> The Poet is a kinsman in the clouds
> Who scoffs at archers, loves a stormy day;
> But on the ground, among the hooting crowds,
> He cannot walk, his wings are in the way.
>
> (Baudelaire 1993, 15–16)

Tristan Corbière in his *Wry-Blue Loves* (which happened to include 116 different animals altogether) turned the traditional romantic bestiary inside out. While Symbolism exalted swans, pelicans, or nightingales, he addressed the reader through the mouth of an unsightly amphibian: "Look at him, a shaven poet, without wings, a nightingale of mud . . .— Horror! [. . .] he returns, cold, under its stone. . . . Good evening—this toad, it's me" (Corbière 1873, 85). In "The Coastal Robber," he included himself in a hybrid genetic line of sea monsters: "My father was an old thief, My mother an old octopus" (Corbière 1873, 265). Jules Laforgue, too, took pleasure in his self-stylization as typically Decadent molluscs: "Oh! What am I doing here, in this room! Poetry. So what? I'm repulsive, I'm a sordid slug!" (Laforgue 1885, 37).

Incomparably more horrifying, Lautréamont conceived of his most famous work, *The Songs of Maldoror*, as a "confession of a man who remembers having lived fifty years in the form of a shark" (Lautréamont 1869, 222), and who is still able to copulate with a female shark. Among the most precious items in Maldoror's bestiary, the proud creatures that the hero intends to "surpass by his congenital cruelty" include a snake, an amphibian, and four sea animals: "On earth, the viper, the huge eye of a toad, the tiger, the elephant, in the sea, the whale, the shark, the hammerhead, the monstrous ray, the tooth of a seal" (Lautréamont 1869, 33).

Even Apollinaire compared himself to a devilfish:

> Squirting his ink to the sky,
> Sucking his lovers' blood,
> Finding the taste of it good,
> This inhuman monster is I.
> (Apollinaire 2011, 41)

An albatross fallen from the sky, a toad, an octopus, a slug, or a shark—Decadent authors would rather identify themselves (if not artists in general) with creatures traditionally considered to be frightening, repellent, or simply ridiculous. From objects of contempt, animals become, in their eyes, accomplices, silent brothers of the "accursed poet," as misunderstood and rejected by society as they are themselves.

It would, of course, be far-fetched to claim that the Decadent authors "actually put themselves in the skin" of the animals to which they compared themselves. However, a new kind of solidarity between species emerged which overcame the traditional human-animal opposition: faced with a choice between the bourgeois conventions and the "disgusting" animal, the Decadent resolutely opted for the latter.[6]

Moreover, behind what traditional sensibility perceived as ugliness or disability, the accursed poet seemed to discover a sort of fragile beauty. On land, the wings of the albatross do not "serve" any purpose; they even make a mockery of their owner, but they are nevertheless what enables it to fly in the skies. Deformity in the eyes of common men was a beauty of a higher order for the Decadents. In the same way, despised diseases—neuroses, epilepsy, hysteria, schizophrenia—implied, in the eyes of the accursed poets, a superior intelligence and sensibility, inaccessible to ordinary mortals. Not having a physical deformity, just like not being sick, became "vulgar." Paul Bourde, a contemporary journalist and playwright, of course exaggerated when he wrote in his pamphlet article about Decadent poets:

> As long as health is essentially vulgar and good for boors, he [the decadent poet] must be at least neuropathic. If blind nature insists on making circulate a banal vigorous blood in his veins, he uses Pravaz's syringe to obtain the morbid condition that suits him.
>
> (Bourde 1889, 24)

Nevertheless, many academic studies have since focused on this "claim of the disease" by Decadent authors.[7]

4. Animals as Models of Perception or Literary Heroes

At the same time, self-stylization as invertebrates, amphibians, or other unusual animals (bats) also implied some sort of experimentation with different organs of perception. *Fin des siècle* writers often endeavoured to grasp reality differently, through the perception of these animals.

For instance, as Evanghélia Stead emphasizes, toads and frogs were as omnipresent in Jean Lorrain's work as in his study room. The Decadent poet never stopped questioning the eyes, the skin, the movements, the amphibians' adaptation to the environment, which he considered an "ugly Narcissus" (Stead 2005, 143), and which he ended up identifying with. More than that, the sight of a blind toad taught the young poet pity and reconciled him in a certain way with the Christian imagination. The symbolic walks here, hand in hand, with the zoological (Stead 2005, 147).

For his part, Robert de Montesquiou[8] remained convinced that the life of his family was for centuries fashioned by the Courtanvaux Castle, a place wrapped in a disturbing atmosphere and conducive to supernatural manifestations, where his ancestors lived amidst clouds of bats. Equipped with a sixth sense, like the sonar of these chiroptera, the poet adopted the "ideogrammatic silhouette of the bat standing on the moon" (Bertrand 1996, 155) to create a personal stamp on his envelopes and stationery. His most famous collection of poems is called *The Bats* (1892).

Decadent aesthetics would also be unthinkable without helminths and segmented worms. Many authors developed the absurd traditional Baroque *memento mori* scenes, turning worms—originally representing moral lessons about the inevitability of death—into real heroes of morbid mini-epics taking place at lovers' tombs:

> Worms will devour us and from my lips they will move over to your lips, creeping along our arms as they like. Our bodies will merge into one. Joined so that no earthquake separates us, we will fall prey to helminths.
>
> (Champsaur 1882, 364)

Death, or at least hibernation, was the basis of many metaphors involving animals: for example, Huysmans' des Esseintes in *Against Nature* lives "on himself, feeding on his own substance, like those torpid creatures that bury themselves away in a hole all winter long" (Huysmans 1998, 62). He even forces the flowers he likes to surround himself with to "copy the internal membranes of animals, to borrow the living colours of their rotting flesh, the magnificent horrors of their gangrened limbs" (Huysmans 1998, 77), as real nature scares him with its endless banality.

In the Decadent world where the artificial was preferred to the natural, the sickly to the healthy, the strange to the common, animals became the basis of unusual and bizarre metaphors: the solitary poet lived his life in a hole, like hibernating mammals, and only tolerated around him flowers evoking animal bodies. The fauna thus became a kind of labyrinth of flesh, a castle haunted by butcher phantasms. Death (received as well as given) was feared but simultaneously fascinated, and its companion the helminth was promoted to the rank of a real actor in the stories. For the first time in the history of literature, the authors of the *fin de siècle* were interested in helminths as such (Palacio 2011, 47), outside of philosophical, religious, or moral metaphors.

Let us take the example of the tapeworm: Léon Hennique, Naturalist and disciple of Zola, chose this botriocephalus as "the protagonist" of a short story named "Benjamin Rozes." The tapeworm leads the life of an "inconvenient tenant" in the intestines of a retired notary. It lodges itself in the most secret intimacy of its host, and its length becomes an object of obsession and nightmares for Mr. Rozes. Alain Montandon mentions in this regard a "Naturalist narrative that does not hesitate to dive into the excrement raised to the height of tragedy" (Roman and Tomiche 2001, 177).

The Decadents adopted this idea of a tenia hero, and introduced it into the "Symphony in minor green" (Symphonie en vert mineur), which formed a part of their famous parody *Deliquescences of Adorée Floupette*:

> The Tenias
> That you denied
> Traitorously got up and left.
>
> (Vicaire and Beauclair 1923, 64)

In *A Question of The Heart*, Gustave Ghèz went so far as to imagine some heroic (on the side of the tapeworm) and therapeutic (on the side of human patients) poems that valued the anatomy of the fearsome worm:

> There is a more terrible and fierce tapeworm:
> It is the one whose head, with its voracious mouth,
> wears a fatal crampon! Armed tapeworms
> Take root in hidden recesses of
> affected stomachs that their presence embarrasses.
> (Ghèz 1888, 180)

The presence of the worm implied neither a social satire of parasitism nor a metaphor of the Decadent poet. In Gustave Ghèz's text the tapeworm is . . . a simple tapeworm. A cestoda living in the digestive tract of vertebrates.

5. Animals in Sadean Experiments and Social Revolts

A unique position in the Decadent bestiary was held by animals designated for various experiments. Decadence revelled in the systematic transformation of plants and animals into works of art, but one of the unwritten rules of these aesthetic orgies was that the "object" rarely survived its metamorphosis. The aforementioned Huysmans hero had been looking so long for a moving thing that would sufficiently "blur the vividness" (Huysmans 1998, 35) of his carpet in the dining room that he ended up with a turtle, whose carapace he covered with gold and encrusted with a mixture of real and artificial precious stones. The poor victim of his refined taste unfortunately soon dies, because "it had not been able to bear the dazzling splendour thrust upon it, the glittering cape in which it had been garbed, the gems with which its back had been encrusted, like a ciborium" (Huysmans 1998, 43).

A similar fate was described in Villiers' "The Legend of the White Elephant." An albino elephant is first intricately painted to be secretly exported from Burma and subsequently bleached several times without success. In the end, the animal is overstuffed to death with aphrodisiacs supposed to ensure the transfer of its precious genetic deviation to further generations.

The category of animal martyrs of beauty also included the swans mentioned in another parodic story called "The Swan-Killer" by the same author, in which a bored, snobby townie strangles these noble birds with iron gloves so that he can enjoy "the strange sweetness of timbre" of their last song (Villiers de l'Isle-Adam 1986b, 136).

Such cynicism might be surprising in the case of writers who, in their daily lives, mostly campaigned for the animal cause, as will be shown in the following. However, let us not forget that these were ferocious parodies of contemporary society and, in particular, of its most repulsive component in the eyes of the Decadents: the petty bourgeois. Faithful to Baudelaire's teaching that "man bites with laughter" (Baudelaire 1995, 528), Villiers de l'Isle-Adam developed a specific literary genre of the "cruel tale," which deployed the most varied modalities of an ultra-aggressive and critical comic against the values of ambient positivism: in a letter to Stéphane Mallarmé on 11 September 1866, the author asserted that "[t]he fact is that I will do to the bourgeois, if God gives me life, what Voltaire did to the clergymen, Rousseau to the aristocrats, and Molière to the doctors" (Villiers de l'Isle-Adam 1986a, 99). The use of vivid descriptions of animal tortures was only one of the way "*fin de siècle* bourgeois sadism was portrayed" (Sylvos 1991, 75), which was also targeted at women, children, the sick, beggars, prisoners sentenced to death, and even corpses. The borderlines between satire and dark humour are rather uncertain in Villiers' texts, so that cruel tales plunge the reader (both contemporary and modern) into a deep malaise.

Other animals worth mentioning include those used by Decadent heroes to express their revolt, or which are literally hurled at a despised mankind. Des Esseintes takes retrospective revenge on his parents by thinking of them while sexually frolicking with prostitutes. There is an involuntary witness, a sort of symbol of their post-mortem presence in their son's house: a captured cricket.

> Thus, out of loathing and contempt for his childhood, he had hung from the ceiling of this room a little cage of silver wire, in which a captive cricket sang, just as crickets had sung among the cinders on the hearths of the Château de Lourps; listening to that song he had heard so often, all the constrained, silent evenings he had spent with his mother, all he neglect he had experienced during his sickly, repressed youth surged up within him, and then, aware of the paroxysm of the woman he was mechanically caressing, whose words or laughter broke into his vision and brought him abruptly back to reality, to the bedroom, to earth, his soul would be filled with turmoil, with a need for revenge for the miseries he had endured, with a fierce urge to defile the family mementoes, with a furious desire to lie, panting, upon cushions of human flesh, to explore to the last drop the fiercest and the bitterest of carnal excesses.
>
> (Huysmans 1998, 10–11)

In his short story called "Christians to the lions!" (*Extraordinary Tales*), Villiers de l'Isle-Adam introduced himself as an ironic spokesman of a group of lions who challenge the Society for the Protection of Animals. They complain of the constraints and sufferings imposed on them by the tamer of the Pastoral Folies (narrow cages, fire-burns, revolver shots, stupid and distracted audiences, poorly executed music). After enjoying a short truce—due to the wounds inflicted upon the tamer by one of the lionesses—four lions are now to "play" in the theatre in a fairy tale. The narrator finds animal torture in laboratories acceptable in the name of science—

> That a Claude Bernard exercises his severity (required by science) on domestic or ferocious mammals (and even makes them previously voiceless—so that their cries, torn by experimental research, do not disturb, in the neighborhood, the peaceful sleep of the townspeople), it is, without a doubt, a criminal necessity.
>
> (Villiers de l'Isle-Adam 1970, 201–2)

—but he is rebelling against abuses imposed on lions by bourgeois art in the middle of Paris, for the pleasure of families with children. It requires worthier and better living conditions for the animals, corresponding to their natural environment, before concluding with a provocation: "After careful consideration (and, especially, given the solemn occasion of yesterday, September 4 [e.g. on the day of the *proclamation of the Third Republic in France—EVB*]), I require, Mr. Chairman, their pure and simple release" (Villiers de l'Isle-Adam 1970, 204).

Thus, the lions would be released into the streets, and Parisians thrown to them. Claiming the same rights and justice for all species, Villiers attacks here (as part of a satire, of course) the anthropocentric paradigm, as well as the idea of human superiority over animals: the lion is neither a toy, a model, nor a theatre decoration.

The same tendency to play with the hierarchy of species can be found in the "mots d'esprit" of Alphonse Allais, who proposed that microbes or plants be protected as much as men or animals, if he did not see them already associated in unions:

> Hygiene is fashionable, microbes are doing worse and worse. And the Society for the Protection of Animals does not make any move!
>
> (Allais and Chouard 1996, 387)

> I founded the Society for the Protection of Plants. We are laying mattresses under the trees to cushion the fall of the leaves.
>
> (Allais 2016, 66)

> • Damn, it must be a little annoying [for the microbes to kill human beings].
> • Annoying! I see you do not know microbes. . . . It just makes them shrug.
>
> The poor tuberculous man is killed on the spot. He raises his eyes to the sky, trying hard to visualize the image of an Union of microbes shrugging their shoulders.
>
> (Allais 2016, 886)

In a much more violent and dark register, Lautréamont imagined in *The Songs of Maldoror* a real veneration of the louse, a royal creature who "would, if its legitimate needs were not satisfied, be capable, by means of an occult power, of becoming big as an elephant and crushing men like ears of corn" (Lautréamont 1869, 92). The author invites mankind to be voluntarily eaten by these insects, which are infinitely superior to them in terms of strength and willpower:

> You may be sure that if their jaws conformed to the measure of their infinite desires, your brain, the retinas of your eyes, your spinal column and all your body would be consumed. Like a drop of water. Take a microscope and examine a louse at work on a beggar's head; you will be surprised. [. . .] But woe to the sperm-whale that fought against a louse! Despite his size, he would be devoured in a trice. Not even his tail would remain to tell the news. An elephant can be stroked. But not a louse.
>
> (Lautréamont 1869, 93–4)

As we can see, despite its proclaimed hostility, or at least scepticism, towards everything non-artificial ("Nature is a grand lady to whom I would like to be introduced, because everybody talks about her and nobody has ever seen her!" (Villiers de l'Isle-Adam 1986b, 832)), Decadence used animal metaphors or identifications very often and with pleasure, not only to shock the bourgeois, defend new aesthetics, or explore the mysteries of hybridity but also to challenge human supremacy, analyse the perceptive experience of animals, or to defend them against the cruelty of modern societies. Decadent self-projections into animals still remained, of course, very marked by Romanticism and anthropocentrism, but some elements of a kind of new solidarity between the "man (of letters) and animal" had begun to appear.

6. The "Philoxenic" Heritage of Decadence

In conclusion, we must return briefly to the legacy of French Naturalism in order to assess its alteration by the Decadents. In his preface to *Therese Raquin*, his first "physiological novel," Émile Zola sums up his conviction that truth must be adhered to and the human body must be studied without regard to social conventions or any emotional reactions of the public. The writer must be like a scientist, who describes with detachment and impartiality the natural phenomena, including the functioning of human beings:

> In *Therese Raquin*, I have sought to study temperaments and not characters. . . . Therese and Laurent are human brutes, nothing more. . . . The amours of my hero and heroine are the satisfying of a

> necessity; the murder they commit is a consequence of their adultery, a consequence which they accept like wolves accept the slaughtering of sheep; finally, that which I have been obliged to term their remorse, consists in a simple organic disorder, in the rebellion of a nervous system strung to the point of breaking. The soul is entirely wanting; I admit this the more readily as I wished it to be so. [. . .] While engaged in writing *Therese Raquin*, I forgot the world, I became lost in the minute and exact copy of life. . . . [I am only] a mere analyst, who may have forgotten himself amidst human putrefaction, but who has forgotten himself there like the doctor forgets himself in the dissecting-room.
>
> (Zola 1883, II–III)

The Naturalists started from the presupposition of Descartes that what we have in common with animals is precisely the body as a machine: in their eyes, even the behaviour of humans can be fully explained through their physiology, and no longer through a rational and immortal soul, distinct from the body.[9]

The Decadents pretended to accept this equation "man = animal," but they pushed it to its most absurd extremes and thus diverted it from the Naturalist logics. With a good deal of humour, the writers of the end of the century identified themselves with the animals "least accepted" by the society of their times and turned them into their accomplices in their iconoclastic strategies. Monkeys or dogs, snails or worms, sharks or devil fish, everything could be used and thrown in the face of the hated bourgeois.

Sarcasm, however, is never completely "innocent." There are many reasons to believe that in comparing themselves with animals, the Decadents were deeply concerned about the fate reserved for the latter. Baudelaire had already felt a secret complicity between street cats and poor people. Zola became a vegetarian out of love for animals (like Leon Tolstoy, Mark Twain, Franz Kafka, George Bernard Shaw, H.G. Wells, Isaac Bashevis-Singer, Marguerite Yourcenar, and many other writers). Charles Cros and Villiers de l'Isle Adam adopted a dog they considered to be the true "reincarnation of Charles Baudelaire." Huysmans was a cat lover, while most Decadents were surrounded not only by dogs but also by exotic animals and birds.

It is no coincidence that the second half of the nineteenth century saw an extraordinary development of the Society for the Protection of Animals, founded in 1845. On 2 July 1850, general de Gramont enforced a law on the ill-treatment of domestic animals. In 1892, the concept of animal rights became the theme of an influential book, *Animals' Rights: Considered in Relation to Social Progress*, by the English social reformer Henry Salt. By the beginning of the twentieth century, associations defending animal welfare and laws against cruelty to animals were present in almost all countries of the world. Among the authors of the time who espoused the animal cause and became activists in these kinds of associations, we should mention Théophile Gautier, Villiers de l'Isle-Adam, Émile Zola, Paul Valéry, Jean Cocteau, and Colette.

Symbolism and Decadence, therefore, are contemporaneous with the first modern wave challenging human exclusivity. In their desire to found a new poetic, and radically opposed to the positivist and naturalistic tradition, late-nineteenth-century writers helped to "rehabilitate" certain animals (the recognition of an accursed poet usually went hand in hand with a new look at his animal avatars) and blurred the boundary between man and animal substantially more than previous literary currents. In short, the famous "contempt for nature," which the Decadents proclaimed loud and clear generally, only rhymes with a certain rejection of cruelty and vulgarity in favour of a more refined and more liveable artifice. It was never a question of denouncing or belittling the animal for the benefit of man. Most of the time, it was quite the opposite.

Notes

1. This chapter is included in the research project PROGRES Q14 "Crisis of Rationality in Modern Thought" and was supported by the European Regional Development Fund-Project "Creativity and Adaptability as Conditions of the Success of Europe in an Interrelated World" (No. CZ.02.1.01/0.0/0.0/16_019/0000734).
2. See for example Baudelaire (2013) or Verlaine (2016).
3. All quotes are translated by the author, unless indicated otherwise in the bibliography.
4. The cabarets of *fin de siècle* Paris gave birth to dozens of "dissident" literary groups where humor blended with blasphemy and political provocation. The Zutists (from the interjection "zut!," "damn!") was an informal club of poets, painters, and musicians who used to meet at the Hôtel des Étrangers in Paris from September to October 1871. The group's meetings resulted in one *Zutist Album*, a sort of experimental poetic logbook, in which the members of the group ferociously caricatured Parnassian poets through parodic texts and provocative drawings. The Hydropaths (etymologically "those that water makes sick") was created in 1878 by the poet Émile Goudeau, who chose this name perhaps inspired by Joseph Gungl's waltz called *Die Hydropathen*. It may be also a pun on the name of the founder (since Goudeau can be read as "goût d'eau" or "goutte d'eau," which means "taste of water" or "water drop"), or an ironic nod to the hydra, the animal whose heads grow back as they are cut off.

   Regarding the other groups, we should mention at least the Hirsutes ("hirsute" meaning "ruffled" or "tousled"), active from 1881 to 1884, and the Jemenfoutistes, who owe their name to the phrase "Je m'en fous" ("I don't care").
5. For the abundant intertextuality of the *fin de siècle* novels, and the various borrowing strategies used by the authors, see at least Montandon (2001).
6. At least, such was the thesis defended, for example, by Camille Lemonnier in *The Toad's Child*, a surprising collection of Naturalist-Decadent short stories, which led the author to trial in 1888 for moral offense. Lemonnier described mineworkers living in atrocious conditions and behaving like animals. At the

end of the novel, the heroes inseminate collectively a revolted widow who hopes to give birth to a social Messiah. A powerful toad's son—"a child of the whole pack"—will finally challenge bourgeois society! (Lemonnier 2015, 126)

7. Among the best known, we should mention Thorel-Cailleteau (2000) and de Palacio (2011).
8. Robert de Montesquiou (1855–1921) was a French aesthete, Symbolist poet, and art collector. He is reputed to have inspired the characters Jean des Esseintes in Joris-Karl Huysmans's *Against Nature* (1884), and Baron de Charlus in Proust's *In Search of Lost Time* (1913–1927).
9. At least in theory, because their actual novels were not devoid of strong bourgeois moralization, on the one hand, and of humanistic or socialist ideals, on the other.

References

Agamben, Giorgio (2004). *The Open: Man and Animal.* Translated by Kevin Atell. Stanford: Stanford University Press.

Alexis, Paul (1971). "Naturalisme pas mort." In *Lettres inédites d'Émile Zola à Paul Alexis 1871–1900.* Toronto: University of Toronto Press.

Allais, Alphonse (2016). *Œuvres complètes.* Paris: Arvensa Éditions.

Allais, Alphonse, and Chouard, Robert (1996). *Alphonse Allais ou L'Humour français.* Paris: Éditions du Rocher.

Apollinaire, Guillaume (2011). *The Bestiary, or Procession of Orpheus.* Translated by X. J. Kennedy. Baltimore: Johns Hopkins University Press.

Baudelaire, Charles (1993). *The Flowers of Evil.* A new translation by James McGowan with parallel French text. Oxford: Oxford University Press.

Baudelaire, Charles (1995). "De l'essence du rire." In *Œuvres complètes*, edited by Cl. Pichois. Vol. 2. Paris: Gallimard, Bibliothèque de la Pléiade.

Baudelaire, Charles (2013). *Notes nouvelles sur Edgar Poe.* Paris: Bibebook.

Bentzon, Thérèse (1889). *Tentée.* Paris: Calmann Lévy.

Bertrand, Antoine (1996). *Les Curiosités esthétiques de Robert de Montesquiou.* Genève: Droz.

Bourde, Paul (1889). "Les Décadents." In *Les Premières Armes du symbolisme.* Paris: Léon Vanier.

Champsaur, Félicien (1882). *Dinah Samuel.* Paris: Ollendorf.

Clerget, Fernand (1891). *Henri Pivert.* Paris: Genonceaux.

Compain, Mélanie-Louise (1903). *L'Un vers l'autre.* Paris: Stock.

Corbière, Tristan (1873). *Les Amours jaunes.* Paris: Glady Frères.

de Goncourt, Edmond, and de Goncourt, Jules (1864). *Journal des Goncourt. Mémoires de la vie littéraire.* Paris: G. Charpentier.

Ghèz, Gustave (1888). *La Question du cœur.* Paris: Calmann Lévy.

Huysmans, Joris-Karl (1998). *Against Nature.* A new translation by Margaret Mauldon. Oxford: Oxford University Press.

Laforgue, Jules (1885). *Complaintes.* Paris: Léon Vanier.

Laforgue, Jules (1891). *Les Amours jaunes.* Paris: Léon Vanier.

Laurent, Émile (1897). *La Poésie décadente devant la science psychiatrique.* Paris: Maloine.

Lautréamont, Isidore Lucien Ducasse, comte de (1869). *Les Chants de Maldoror.* Paris/Brussels: Publisher Non Identified.

Lemonnier, Camille (2015). *L'Enfant du Crapaud.* Paris: Arbre vengeur.

Millaud, Albert (1887). *Physiologies parisiennes*. Paris: La Librairie Illustrée.

Montandon, Alain (2001). *Mythes de la décadence*. Clermont-Ferrand: Presses Universitaires Blaise Pascal.

Palacio, Jean de (2011). *La Décadence. Le mot et la chose*. Paris: Les Belles Lettres.

Roman, Myriam, and Tomiche, Anne (2001). *Figures du parasite*. Clermont-Ferrand: Presses universitaires Blaise Pascal.

Rosny, J.-H. (1890). *Le Termite*. Paris: Savine.

Stead, Évanghélia (2005). "Monstres à facettes, images de soi: la mandragore et la grenouille dans *La Mandragore* de Jean Lorrain." In *Particularités physiques et marginalité dans la littérature*, edited by Arlette Bouloumié, 141–50. Rennes: Presses universitaires de Rennes.

Sylvos, Françoise (1991). "L'Essence cruelle du rire: Villiers de l'Isle-Adam." *Romantisme* 74: 73–82.

Thorel-Cailleteau, Sylvie (2000). *Dieu, la chair et les livres. Une approche de la décadence*. Paris: Honoré Champion.

Trézenik, Léo (1887). *La Jupe*. Paris: Monnier.

Verlaine, Paul (2016). *Les Poètes maudits: Tristan Corbière, Arthur Rimbaud, Stéphane Mallarmé*. Paris: Hachette Livre BNF.

Vicaire, Gabriel, and Beauclair, Henri (1923). *Les Déliquescences d'Adoré Floupette*. Paris: Jonquières.

Villiers de l'Isle-Adam (1970). *Œuvres complètes VI. "Histoires insolites"*. Genève: Slatkine Reprints.

Villiers de l'Isle-Adam (1986a). *Correspondance générale et documents inédits*. Vol. 1. Édited by J. Bollery. Paris. Mercure de France.

Villiers de l'Isle-Adam (1986b). *Œuvres complètes*. Vol. 1. Paris: Gallimard, Bibliothèque de la Pléiade.

Zola, Émile (2010). *Correspondance*. Vol. 11. Montréal: Presses de l'Université de Montréal.

Zola, Émile (1883). *Thérèse Raquin suivi du Capitaine Burle*. Préface à la deuxième édition. Paris: Flammarion.

Part III

"Across the Narrow Abyss of Non-comprehension"

Entering Animals' Lives

11 Literature and Animal Expressiveness

On the Cognitive and Ethical Aspects of Zoopoetics

Anne Simon

Zoopoetics, which I have been working on shaping for 20 years, aims to highlight the plurality of stylistic, linguistic, and narrative media used by writers to express the plurality of animal actions, affects, and worlds. Zoopoetics goes even further, as it considers as impossible the exclusive focus on animals: it therefore underlines the intricacies of their interactions with other animals (especially human beings), plants, wind, minerals, and what was long ago called "elements."[1] Such an approach is based on a renewal of disciplinary interfaces. If the dialogue with philosophy, ethics, or history needs to continue, I see it necessary to tackle more unusual overlaps, such as between literary studies and legal studies, anthropological perspectivism, religious studies, mesology, zoosemiotics, or ethology. The issue is obviously not for one to become an expert in all of these disciplines but to understand what they discuss when they place animals at the core of their investigations. In other words: what, outside literary studies, operates as a stake, a catch, a blind spot, or a driving force? Finding out what is at stake in another field allows one to shift focus, shake up one's routine, and reshape the approach of one's own field of research. It also enables one to reveal the potential contributions of literary studies to the animal question, and to the broader question of life.

Focusing on other ways of thinking is very fruitful, as it leads one to question anew the narrative and syntactic structures, the rotations of rhythms and points of view, the thematic innovations that a writer brings into play when writing about animality, even laboriously. Indeed, writers' ways of writing enable them to engage their reader in the world of a specific animal while admitting the extent to which this world eludes them! In both cases, the relationship between literary creation and the life of an animal seems to be built on a paradox. Certainly, nothing is more specifically human than the creative language based on imagination, e.g. literature, which Merleau-Ponty characterized as the "language to the second power" (1969, 7, 132). It seems to be the pinnacle of language, and in some way, of man (from a temporary logocentric perspective). More than the most astonishing abstract demonstration, its poetic and fictional application brings the language to its higher degree of figurality, polysemy, and complexity. As Lessing would put it, the novel, with its intrinsic tie to its characters' history, actions, and movements, can be viewed as the art of succession in time. It is in fact the ideal genre, as it connects the art of time and the exploration of places, growing rich through bifurcations, red herrings, and detours. The "scattered great fugue" of animals, broached by the French writer Jean-Christophe Bailly (2011, 109), reveals to us the fascinating plurality of life forms. Many writers are guided by this fugue into quirky—typically non-human—ways to enter linguistic forms and narrative composition. No wonder that many writers who trailed an animal strove to plunge "Into the Unknown's depths, to find the new" (as Baudelaire wrote in his poem "The Voyage").

1. Animal Poetics and Biomorphism

Such profusion and dissemination of the novel's discourse, of which *Moby Dick*, in my view, is the paradigmatic case, can in turn portray animals as voiceless beings, "for which" and "in place of which" writers must speak (cfr. Deleuze 1995)—writers, but also philosophers who consider themselves as distraught animals "on the lookout" (Deleuze 1995). But the very possibility of such assignment shows that animals are not actually mere *aloga*: they continue carrying out communication strategies, passing on their patterns from generation to generation or, like potential novelists, "tell stories." They are, for instance, capable of deceiving their equals or humans—as we will see in the case of the old deer fooling blundering hunters in French writer Maurice Genevoix's *La Dernière Harde* (*The Last Herd*) (1988, 98–102). It is a known fact that shared spaces of meaning always turn up among such "mixed communities," as highlighted by Arne Naess, Mary Midgley, Jocelyne Porcher, Dominique Lestel, and Vinciane Despret, and it should not come as a surprise. Indeed, according to many linguists and philosophers, parts of human languages originated in animal forms, such as those related to body parts, or types of locomotion or noise emissions. Just think of onomatopoeia and the many other words whose meaning is conveyed by their sounding, such as the *hiss* of the snake.[2] Or if we look into scripts, take Egyptian and Asian ideograms, or the Hebrew consonant *gimel* (the equivalent of our *g*), which means "camel" and reminds us of a camel's head and front legs (see Abram 1996, 100–1).

"You have been the first ciphered signs of the written language, the first stenography of thought, and the epitaphs of the very first skies ever contemplated by a man who could not be but illiterate," wrote Ramon Gomez de la Serna in his letter to the swallows, which, in the sky, "are like the invention of the printing press" (2006, 27, 30). Animals are thus more figurative or rhetorical than is usually thought. Their relationship to the world, even guided by instinct, does not conflict with expression nor with invention: agency (*poïen*) is constitutive of their existence, from the most basic organisms on. The ability to act upon the environment and to respond in a more adapted, if not more innovative, way to its unsteady and undetermined nature is not limited to humans. Supporting his assumption with the studies of ethologists, linguists, and philosophers, Aaron M. Moe has recalled how animals can create bodily forms of responses (through sounds, gestures, behaviours, rhythms, and so forth) to the requests of others and of the world, whether to communicate within or beyond their species, to lay traps, to make sense of their environment, or to act upon it (Moe 2014, 5–32). Primordial poetics of the living—a *poïen* of the *zôon*—thus exists, which is also a pre-literary way to define zoopoetics. One can understand that many writers consider their creation as the expression, if not the foundation, of an ontological relationship between living and writing.

Humans' illusory self-constitution, as if based on a linguistic mastery enabling them to extract themselves from "the lot of other creatures" (de Montaigne 1988, 12), should therefore be discarded. Such a gesture of classification, repartition, and distribution, which is actually a global gesture of deprivation and separation, has since Montaigne been decried as misguided and abusive by philosophers sensitive to the expressive capacity of living beings. No wonder that Derrida, the first philosopher who coined the notion of "zoopoetics" (Derrida 2006, 20), used this concept to refer to Kafka, whose diary and many short stories were dedicated to the psychic and political fluxing of the human/animal divide.[3]

But let us go back to Deleuze, Guattari, and Merleau-Ponty, who believe that creative language enables, if not the transformation into, at least an approach and encounter with these other beings closest to us. To speak "in place of" animals: Deleuze's phrase shall be construed in its plentiful, carnal, re-routing meaning—"for," "as," but also "in the territory of," "in the world of," if not "with the body of." For even though "becoming a dog" is not possible, one can "feel at one with the animal, one entity with no organs defined by zones of intensity or zones of proximity" (Deleuze and Guattari 1980, 335). The author's attempts at zoomorphing, by crawling into an animal's skin only then to shed that skin, shows the writers' ability to divert from their expectations, to dwell differently, to experiment a new *oikos*. The novel has sometimes tread the path of deadly mixtures between men and animals: think of *The Island of Doctor Moreau* by H.G. Wells, or *The Lion* by Joseph Kessel, a so-called children's book whose plot is actually based on a bestial, quite erotic relationship between a young teenage girl and a lion![4] More interestingly, in her sleep, Béatrix Beck mixes her breath with that of her female cat and dreams of limbic pairings: "Last night found myself breathing beside myself. It was her, on the pillow" (2007, 36).[5] In any case, writers consider the novel as a carnal, spatial, and mental venture of self-disorientation.

Focusing on twentieth- and twenty-first-century novels, I would like to show henceforth that novel-writing lies on several psychic positional and narrative processes which lead humans to expand their points of view, even to undertake others'. Novels can leave behind the opposing concepts of animality and humanity to enter into animal worlds, whose linguistic characterization, less paradoxically than it seems, enables us to take part in a populous world. Language opens us up to alterity, whether it be relative (like the dog suffering like a human in Romain Gary's work or the mule trudging through our history in Vassili Grosman's "The Road") or radical (like the odd, absurd, and wonderfully baffling sea cucumber so dear to Frédérick Tristan).

2. Animal Exoticisms and Novelistic Perspectivism

Gazing at his cat, who was actually a special female cat, Derrida found himself confronted with the aporia of philosophizing in a particular way: "what we have here is an existence that defies any conceptualization" (Derrida 2006, 23). Jean-Christophe Bailly goes further, saying that the concept of "'animality', which is non-existent," should be dropped and replaced by the imaginary experience of self-displacement (Bailly 2013, 75). "An existence," stressed Derrida in the earlier quote: exploring the novel, a genre of specificities and situations, of biographies and personal stories, compels one to move from the universality of animality to the plural diversity of animals—to what in the Bible is called "multiplication" and "fruitfulness"—as well as to their individuality. For some, such novelistic animals may be viewed as sheer "*animots*" (animal-words), as Derrida coined, e.g. human, all-too-human creatures. Such a position obliterates the fact that the word, from a post-phenomenological perspective (think of Merleau-Ponty), has to do with experience and the shift into existence. And what an experience! That of plural, differentiated animals, which cannot be assigned to specific automatisms, which are the dream, or rather the nightmare, constructed by laboratory ethology in such diverse works as Daniel Keyes' *Flowers for Algernon*, Malaparte's *The Skin*, or Olivia Rosenthal's *What Do the Reindeers Do After Christmas?* (*Que font les rennes après Noël?*).

Pergaud, a writer who was also a long-gone type of hunter, put it thusly: a view from above tells us nothing about animals but only refers to a human way of positioning oneself; and that positivism does not tell us anything about animals but refers to a human way to construct them as objects. A social writer and committed humanist, Pergaud would also rail against vivisection and object to "taking part in the mawkish and bleating flock, or in the slaying and whining phalanx" of those exclusively defending animals (Pergaud 1987, 1032). He nevertheless knew that to understand living beings that are unlike us, one needs to abandon the concept of "animality" and launch into the narration of the "The Lives of Animals," the title of his posthumously published collection. To do so, one must play with our relationship with the outer and the inner animal, such a strange familiar sibling: it is necessary to replace the artificial laboratory experimentation with narrative based on an ethological observation, as the entomologist Fabre did before Pergaud. Yet also, at the opposite end, it is necessary to shift from the "overviewing" process—which Merleau-Ponty criticized so much in his lecture on nature, as well as in *The Visible and the Invisible*—to an experience of introjection (see Merleau-Ponty 1964, 1994). That is sheer anthropomorphism, one may contend: a novel is but the result of human imagination, and Pergaud's Margot (the magpie), Genevoix's Red Deer, Melville's Moby Dick (the whale), Kessel's King (the lion), Sorman's Culotte (the cow), and Sorente's Marina (the sow) are but ghosts or fantasies of ourselves, novelist and reader alike.

The problem of animality in the novel compels us to reopen the question of the ontological and exotic power of literature, as explored by Victor Segalen. Let us leave aside the debate of the 1960s, often reduced to the supremacy of signs and a universal agreement around autotelic texts, which have overshadowed the Geneva school, or more broadly, post-phenomenology. As stated previously, if the novel cannot be reduced to the projection of a phantasmagorical world, or to a solipsistic experience of reality, if it does not settle for the exhibition of mere linguistic structures, it is because the language whose articulations the novel displays is not an *ex nihilo* skill or a specificity which would oppose us to animality. Just like dance, which is another paradoxical art in the sense that the human body escapes itself in the most extreme self-proximity (think of the shows of "compagnie du Guetteur," in which Luc Petton directs birds and humans), the literary experience does not antagonize the sensible experience: they resect from each other, and Merleau-Ponty rightly spoke of the correspondence between the "composition" or the "deep syntax" of a narrative and the "lines of forces" of what he calls the "landscape of an experience" (1968, 40).

The question of anthropomorphism, often deemed a fault or an aporia, should therefore be revoked: human beings, supposedly only anthropocentric, may not escape their human skins, and their language taints with humanity everything it touches. As if human language was only capable of speaking about mankind, or could not speak of something other than itself! Are we not denying the cognitive efficiency and profit of anthropomorphism insofar as not all anthropomorphisms are alike? I am hinting at the distinction proposed by Baptiste Morizot between a spontaneous anthropomorphism leading to false interpretations and a heuristic one, which employs analogies answering scientific protocols (2016, 163). Furthermore, the aforementioned benefits of studies on the prehistoric and neolithic co-evolution of human languages and animal sounds or sign patterns should not be overlooked. Finally, there is a risk of forgetting that a human is not "sheerly" anthropomorphic but also primatomorphic, mammomorphic, and, more broadly, biomorphic (see Keeley 2004; Morizot 2016, 55, 184): human language is intertwined with everything that determines a human being. A human being indeed is not *only* a human being or is *more* than what we usually think when we speak of "humankind": our teeth and bones are made of minerals; we live because we are in symbiosis with microbes, virus, bacteria, and because we eat and metabolize other living beings. Many novelists understood this without spelling it out: evolution and biomorphistic logics are such that we intuitively understand other species related to us, we share their emotions and expressions, and *we're able to account for this with (and not despite) specific human means*, such as figurative language (see Darwin 1872).

Mikhail Bakhtin generally defined the genre of the novel as the art of polyphony. If the novel can account for animal worlds and affects, this is precisely because it allows language to express itself in an incredible diversity of viewpoints. Such points of view are themselves partly grounded on the faculty of empathy and on its kin faculty of narratorial projection (see Simon 2014): the "conscious mental simulation of the other's perspective" (Decety 2004, 53) could turn out to be consubstantial with novel creation. Genevoix's statement, 20 years after he wrote *The Last Herd*, that "I have been the Red Deer" (2000, 853) should be taken seriously. By saying this, he also identifies with La Futaie, a character obsessed with this very deer, whose life he saved several times before then hunting it for love:

> Before even starting to hunt him, he would have caught up to him, his desire would have blended them together, made him tremble with him, muddling with him the patient maze of his ruses, forestall his impetus as he doubles back to escape.
>
> (Genevoix 1988, 122)[6]

On a stylistic level, using Gérard Genette and Dorrit Cohn's terminology in *Figures III* (1972) and *Transparent Mind* (1978), the novelist makes it impossible to settle for omniscient narration, internal focalization, or third-person psycho-narration conveying the animal's feelings. The young stag just lost the old deer who guided him (and actually did so only to throw the hunters off his track):

> He looked for him [the old deer], shivering with dread and anger. He remembered now that a large wild shadow had jumped out of the road, with such a high and long leap that he took it for a gust of wind in the tawny foliage of an oak tree.
>
> (Genevoix 1988, 102)

As he refrained from projecting a human psychology onto this other life experience, Genevoix did his best to link the emotions of his animal character to its behaviour, its environment (according to von Uexküll and Augustin Berque), or its body parts: "From time to time, one of the hooves of The Red sunk into a loose hole where his leg got engulfed all of a sudden. He stumbled and panted in rage" (Genevoix 1988, 77–8). As he puts himself in the place of the animal, the novelist withholds a distance, enabling him both to half-open and maintain "secret" his inner life (Genevoix 1988, 122). On this level, Jean-Claude Coquet's enunciative entity, Oswald Ducrot's shade of polyphony, and the POV theory developed by Alain Rabatel (2008, ch. 2) are tremendously valuable to account for animals' inner lives. These authors show that speaking subjects, or speech transcriptions, are useless when it comes to producing a phrase conveying a perception, an affect, a representation, or a thought. As for the sources of the enunciation, they can be ambiguous, even unsolvable: most of the time, readers cannot figure out whether they are feeling at one with an animal or with a narrator projecting himself or herself onto the animal. Many novelists such as Maurice Genevoix, Jean Giono, or Vassili Grossman, who did not feel the need for an "animal dialogue" *à la* La Fontaine, or Colette or Marcel Aymé, therefore suggest that a speechless animal can also be the source of perspective in a text. Such a source of perspective can also be an animal-animal collective, an animal-human collective, or animal-nature collective. Sophie Milcent-Lawson has shown that reported speech and inner monologue used to reproduce animals' individualities are not the preferred linguistic process for French novelists from the first half of the twentieth century:

> The demotion of reported speech for the benefit of the represented thought process and accounts of perception, and the preference granted to indirect free speech, attest to the effort made to turn away from an age-old anthropomorphism that was dominant in the literary representation of animals.
>
> (Milcent-Lawson 2017, 91–106)

Genevoix offers us the mutual fascination between the gaze of the deer and the gaze of the hunter: "The Red . . . looks at the man's eyes. His eyes are wide open. . . . The Man only sees two golden reflections, two clear bottomless glows whose smoothness strokes his face" (Genevoix 1988, 248). Writers—such as Rilke, or Hofmannsthal, for instance—help us approach, if not enter, the animal enigma through a *mise en abyme* of viewpoints and interpretations. They also make us aware of specific, e.g. wilful and active, animal minds. The *em* prefix of the word *empathy* does not highlight an unbreakable confinement inside a monadic human: it suggests that a community of affects and sensitivities could actually *lead towards* another "inside" and bring up an *encounter*. One may recall Derrida's beautiful pages on how beasts were long denied their ability to answer or Georges Bataille's writings on the ability of poetry to put into language the paradox of animal otherness, an intimate part of our humanity:

> Animals open up an attractive and familiar depth for me. This depth, in a way, is known to me: it is mine. And at the same time, it is

> concealed far, far away from me: what deserves to be called "depth" is precisely all that escapes me.
>
> (Bataille 1973, 30–1)

Such paradoxical e-motion allows for the coming out of oneself within the most intimate and trying experience of empathy. Far from being a solipsistic illusion, it turns out to be mainly inter-relational, entitling the "mankind-animality intertwining" broached by Merleau-Ponty in his lecture on nature (1994, 269). On the contrary, some writers, from Proust to Jean Rolin, Jacques Lacarrière, or Marie Darrieussecq, attempt to account for the alterity of animal worlds and observe henceforth a distance deemed an unbridgeable gap: it is always in order to express its fascinating eeriness or to move within another body devoid of a human brain. External focalization is then favoured, as it allows one to stay close to animal bodies and motions, even though at times it appears almost brutal in its cold and impassionate rejection of "sympathizing" with beasts, the familiar-familial animals detested by Deleuze or Sartre. Rilke and Hofmannsthal are therefore closer to the panther and the rat, but such proximity denies any merging, as a tribute paid to these other living beings' alterity.

Applying the technique of multiple points of view to depict animal subjectivities should not been viewed as mere linguistics. First, such a view overlooks the language's philosophical ability to actualize: as shown by Sophie Lawson, when choosing any technique, a writer takes sides when considering animal interiority. Second, literary language should not be denied its almost magical, conjuring ability to refer to what falls beyond it. By conjuring, I mean e-vocation in its incantatory and invocatory meanings: "to call on, to attract," "to make appear by incantations, to resurrect (spirits)," to invoke shadows, "to bring to mind by images, associations of ideas."[7] It is necessary to break away from the mimetic when considering the notion of reference: a metaphor is quick and sharp, is "lively," in Paul Ricœur's words, and it only speaks of a vivid world, of a naturing nature, if poetic language is considered a referential target ("visée référentielle," *The Rule of Metaphor*, 1975). Poetic language is a force of enunciation, not an impossible copy-paste of its referent. As Paul Ricœur reminds us,

> Certainly Greek man was far less quick than we are to identify *phusis* with some inert "given." Perhaps it is because, for him, nature is itself living that *mimêsis* can be not enslaving and that compositional and creative imitation of nature can be possible.
>
> (Ricœur 2004, 48)

Contrary to the assumption of many linguists supporting literary autotelism, it may so happen that the word "dog" bites, and that a book, written or read, bewilders and transforms a life. Malaparte, in *The Skin*, conceived of a narrative "larger than life": his dog Febo, locked in an experimental laboratory, with his belly cut open, and his vocal cords cut. This is a way to evoke, against the hypersignificance of the ancient haruspex, the loss of meaning and the silence of pain during World War II.

3. Mutating Metamorphosis

The attempts of literary perspectivism to make us enter the animal environment (von Uexküll 2010) are not, I must insist, doomed to fail from the start. To render the complexity of such perspectivism implies accounting for two *topoi*, probably the most ancient ones regarding the relationship between animals and language: metamorphosis and hybridity, both universal mythological patterns. They deserve further development. I will only underline their ability as they move from myth to novel, to translate existential attitudes of self-affectualization, which "imply both embodiment and virtuality," according to Philippe Jousset (2007, 138). I would also like to stress that, in modernity, at stake are metaphors that are sketched, started, flowing, in other words, never terminated or set.

Writers do not share anymore Ovid's belief in a definite change of body, species, or even realm, as it is generally the case in the *Metamorphosis*. On the contrary, they believe in moult, in a more-or-less natural, fluid, political, and metaphysical shifting from one specimen to the other. Franz Kafka's human-ape in *A Report to an Academy*, the butchered bull in the industrial abattoirs in Alfred Döblin's *Berlin Alexanderplatz*, the half-crushed beetle in Clarice Lispector's *The Passion According to G.H.* are all animals abused by humans, but they share a common destiny and foreshadow Antoine Volodine's post-exotic characters. Indeed, they evolve in a nightmarish world, where the runaways, the dying, the inmates, and the insects all struggle to survive or to destroy each other until they can hardly be differentiated:

> Camps are for cockroaches and the dead.—For us then, as Gabriella Bruna said.
>
> (Volodine 2002, 210)

Another quote:

> Finally the bestial night could be heard, with the cracking of grim darkness, beasts nibbling at the end of the corridor, and, very close, cockroaches mounting or terminating each other.
>
> (Volodine 2002, 21)

For its part, the narrator of the beautiful *The Land Under Bark* (*Le Pays sous l'écorce*) by Jacques Lacarrière fails to wholly transform. He nevertheless succeeds in leaving—*which is something*—his "hominid" skin. Along social, and not genetic and Darwinian, logics, Marcel Proust's bug-man (*In Search of Lost Time*) wavers between the aristocrat and the vibrating larva; Marie Darieussecq's sow-woman wavers between the whore and the sow; Émile Ajar/Romain Gary's snake-man (*Gros-Câlin*) wavers between the desperate lone man of the city and the python, that is the quintessential animal emblem of avoidance and rebirth. And what to make of the undetermined creature in Kafka's *The Burrow*? The conflicting and muddling levels of this last unfinished short story refer both to its author's threatened and precarious Jewish condition and to the Czech retreat into nationalism in 1920s Prague. They relate as much to the obsession with failure as to the nearing of death, as much to the animal desperate to survive as to the writer who has dug his hole within his writing.

The appeal of hybridation generated an animal fiction akin to what is called in philosophy an anthropological fiction: a fiction telling lies (managing to place oneself in the existential gait of an animal) to get at the truth, a fiction compelling both its writer and reader to *displace* their own access to the world, to change their grounding, their fears, and their joys.

5. Neither Humans Nor Animals Are Extraterrestrials

As a sign of our times, these wanderings, these various gaits, these cross-ties which define the animal mode of being as well as the novel genre are undermined today by the objectification of the living. Many contemporary novels hence adopt a zoopolitical point of view stemming from a tradition of standing up to the aporia of mass-livestock farming, which goes all the way back to Leo Tolstoy, Upton Sinclair, Alfred Döblin, Georges Duhamel, and Pierre Gascar.[8] Such a confrontation is based both on ethic and linguistic concerns, which refer to the fact that the usages of language are ways for mankind to inhabit the world. But the industrial or scientific nomenclatures that turn suffering animals into mechanized chimeras pervert the ever-lasting share of human and animal territories and the very meaning of language. To these industrial thanatopoetics, novelists such as Yves Bichet, Jean-Baptiste Del Amo, Tristan Egolf, Sylvain Tesson, and Joël Egloff have answered with *biopoethics*, aiming to credit anew the discredited and disqualified lives of animals and staffers alike. They reinject unforeseen elements, value, and individuality within the food-processing hell, introducing novelistic grains of sand—such as madness, self-destruction, dereliction, love, liberation-dreams—into the unremitting mechanics that must switch on "the Tool," as the slaughterhouse is called by the characters in Isabelle Sorente's *180 Days* (2013, 460). The industrialization of the living indicates the end of the long domestication process which led the hominid to a hominization that failed to take the time to become humanization—and I am hinting here both at Albert Cohen and Romain Gary. Such humanization can only achieve its full potential if humans surrender their fascination for animal suffering, which is nothing but a mirror of themselves (I think of Hungarian writer Károly Pap's short story "Blood,"[9] where the Jewish child comes to understand that the blood of the geese that fascinates him is forbidden not only because it conceals vital breath but also because he shares it with them). Such humanization can only come true if mankind protects and cherishes what Gary calls its "margin" (1956, back cover): the fringe of alterity, uselessness, congestion, and savagery formed by a herd of elephants (1968).

Humans should not bother to draw an impervious human-animal line: like it or not, they are on the fringe, and such overlap is their luck and salvation. We tend to forget that the sixth day saw all terrestrial creatures come to being, humans included. All life forms are indeed in relationship and dependence with an *archè*—an origin, a reason, a refuge, a dwelling: the Earth.[10] This *archè*, which is also our second and last ark, though perceived through different sensory and motor modalities specific to each species, is nevertheless the common foundation on which all living creatures evolve. It is high-time for humans to stop looking at their habitat from above, as if they were not part and parcel of it.

Notes

1. This chapter is a reworked version of an article published in French: Anne Simon, "La zoopoétique, une approche émergente: le cas du roman." *Revue des Sciences Humaines*, no. 328 (2017): 71–89. My best thanks are due to Clara Royer, for her translation, and to Jessica Harrison Caudry and Chiara Mengozzi, for our linguistic debates.
2. See Shepard (1996, 37) and Krause (2013), who founded biophony and geophony.
3. These include "The Village Schoolmaster," "In our Synagogue," "A Crossbreed," "The Hunter Gracchus," "A Report to an Academy," "The Metamorphosis," "Josephine the Singer, or The Mouse Folk," "The Vulture," "Investigations of a Dog," or "The Burrow," amongst others.
4. "Bestiality," in its original meaning of sexual relations between men and animals.
5. In the original version the sentence is written without the personal pronoun "I": "Cette nuit me suis étonnée de respirer à côté de moi-même. C'était elle, sur l'oreiller."
6. Anne Simon and Clara Royer have chosen to translate the pronoun "lui" by "him" rather than by "it," to evoke the intimate relationship between the human character and the animal character.
7. "Évoquer" in *Dictionnaire historique de la langue française* (1998, 1350–51).
8. For further developments, see Anne Simon, *Une bête entre les lignes. Essai de zoopoétique*. Marseille: Éditions Wildproject (forthcoming in 2020).
9. In *Irgalom* (*Miséricorde*), 1937. I thank Clara Royer for sending me her French translation of this short story.
10. See Husserl (1989). On the ark, see Simon (2018).

References

Abram, David (1996). *The Spell of the Sensuous*. New York: Random House.

Bailly, Jean-Christophe (2011). "Les Animaux conjuguent les verbes en silence." *L'Esprit créateur* 51, no. 4: 106–14.

Bailly, Jean-Christophe (2013). *Le Parti pris des animaux*. Paris: Christian Bourgois.

Bataille, Georges (1973). *Théorie de la religion*. Paris: Gallimard, coll. "Idées".

Beck, Béatrix (2007 [1984]). *L'Enfant chat*. Paris: Arléa.

Cohn, Dorrit (1978). *Transparents Minds: Narrative Modes for Presenting Consciousness in Fiction*. Princeton: Princeton University Press.

Darwin, Charles (1872). *The Expression of the Emotions in Man and Animals*. London: John Murray.

Decety, Jean (2004). "L'empathie est-elle une simulation mentale de la subjectivité d'autrui?" In *L'Empathie*, edited by Alain Berthoz and Gérard Jorland, 53–88. Paris: Odile Jacob.

Deleuze, Gilles (1995). "A Comme Animal." In *L'Abécédaire de Gilles Deleuze*, interview with Claire Parnet. Dir. by Pierre-André Boutang and Michel Pamart, Sodaperaga.

Deleuze, Gilles, and Guattari, Félix (1980). *Mille Plateaux*. Paris: Minuit.

de Montaigne, Michel (1988). "Apologie de Raimond Sebond." In *Essais*, edited by Pierre Villey. Vol. 2, 12, 438–604. Paris: Quadrige/Presses universitaires de France.

Derrida, Jacques (2006). *L'Animal que donc je suis*. Paris: Galilée.

Gary, Romain (1956). *Les Racines du ciel*. Paris: Gallimard.

Gary, Romain (1968). "Lettre à l'éléphant." *Le Figaro Littéraire*, March 4, 10–12.

Genette, Gérard (1972). *Figure III*. Paris: Seuil.

Genevoix, Maurice (1988). *La Dernière Harde*. Paris: Flammarion.

Genevoix, Maurice (2000 [1960]). *Jeux de glaces*. Paris: Omnibus.

Gomez de la Serna, Ramon (2006). "Lettre de la première année." In *Lettres aux hirondelles et à moi-même*, translated from Spanish by Jacques Ancet, 24–32. Marseille: André Dimanche éditeur.

Husserl, Edmund (1989 [1934]). *La Terre ne se meut pas*. Translated by Didier Frank, Jean-François Lavigne, and Dominique Pradelle. Paris: Minuit.

Jousset, Philippe (2007). *Anthropologie du style. Propositions*. Bordeaux: Presses universitaires de Bordeaux.

Keeley, Brian L. (2004). "Anthropomorphism, Primatomorphism, Mammalomorphism: Understanding Cross-Species Comparisons." *Biology and Philosophy*, no. 19: 521–40.

Krause, Bernie (2013). *Le Grand Orchestre animal*. Paris: Flammarion.

Merleau-Ponty, Maurice (1964). *Le Visible et l'Invisible*. Paris: Gallimard, coll. "Tel".

Merleau-Ponty, Maurice (1968). "Le problème de la parole." In *Résumés de cours. Collège de France, 1952–1960*, 33–42. Paris: Gallimard.

Merleau-Ponty, Maurice (1969). *La Prose du monde*. Paris: Gallimard.

Merleau-Ponty, Maurice (1994). *La Nature. Notes. Cours du Collège de France*. Edited by Dominique Séglard. Paris: Seuil, coll. "Traces écrites."

Milcent-Lawson, Sophie (2017). "Zoographies. Traitements linguistique et stylistique du point de vue animal en régime fictionnel." *Revue des Sciences Humaines*, no. 328: 91–106.

Moe, Aaron M. (2014). "Zoopoetics: An Introduction." In *Zoopoetics: Animals and the Making of Poetry*, 5–32. Lanham: Lexington Books.
Morizot, Baptiste (2016). *Les Diplomates. Cohabiter avec les loups sur une autre carte du vivant*. Paris: Éditions Wildproject.
Pergaud, Louis (1987 [1923]). "À propos de la vivisection" [18 June 1914], published posthumously in *La Vie des bêtes, études et nouvelles*. Paris: Mercure de France.
Rabatel, Alain (2008). *Homo narrans. Pour une analyse énonciative et interactionnelle du récit*. Limoges: Éditions Lambert-Lucas.
Ricœur, Paul (2004). *The Rule of Metaphor: The Creation of Meaning in Language*. London/New York: Routledge.
Shepard, Paul (1996). "Bears and People." In *Traces of an Omnivore*. Washington: Island Press.
Simon, Anne (2014). "L'animal entre empathie et échappée (Lacarrière, Darrieussecq, Bailly)." *Figures de l'art*, no. 27: 257–68.
Simon, Anne (2018). "Other Temporalities of Life: Zoopoetics and Animal Perspectivism." *Dibur Literary Journal*, no. 6. https://arcade.stanford.edu/dibur/other-temporalities-life-zoopoetics-and-animal-perspectivism.
Simon, Anne (2019). "Zoopoethics: Literature Challenged by Industrial Livestock Farming." In *Dwellings of Enchantment: Writing and Reenchanting the Earth*, edited by Meillon Bénédicte. New York: Rowman & Littlefield (forthcoming).
Sorente, Isabelle (2013). *180 jours*. Paris: Éditions Jean-Claude Lattès.
Volodine, Antoine (2002). *Dongdog*. Paris: Seuil.
von Uexküll, Jakob (2010 [1956]). *Milieu animal et milieu humain*. Translated by Charles Martin-Fréville. Paris: Payot et Rivages.

12 "Une langue ou une musique inouïe, assez inhumaine . . ."

Narrative Voice and the Question of the Animal

Kári Driscoll

1. The Songs of Mice

In 2015, a study published in the journal *Frontiers in Behavioral Neuroscience* reported that male mice not only sing songs to attract females but that these songs are also syntactically complex and vary according to social context (Chabout et al. 2015). This places mice in the same category as songbirds. The principal difference is that, unlike birds, mice sing at ultrasonic frequencies (30–120 kHz), and so their songs are inaudible to the human ear. This was not the first time singing mice had been in the news. In fact, every few years the media report on a new discovery concerning the songs of mice, most of which build on the pioneering work of Timothy Holy and Zhongsheng Guo, who first established that mouse vocalizations have the characteristics of songs, that is, that they consist of several distinct syllables or phrases "uttered in succession and so related as to form a recognizable sequence or pattern in time" (Holy and Guo 2005, 2178). Evidently, there is something about these studies that captures people's imagination. In 2011, for example, the *Smithsonian Magazine* published a piece on a North Carolina project aiming to record and study these ultrasonic songs, remarking that "the world of rodents, long thought mostly quiet, may be full of songs, broadcast short distances, from one animal to another, songs that we still know very little about" (Dunne 2011).

Some mice appear to sing at a pitch audible to the human ear, though it remains unclear why, and reports of such murine singing mice date back at least to the nineteenth century, well before it was possible to record them. Alfred Brehm, in his compendious *Life of Animals* published in the 1860s, describes the phenomenon in some detail. Some people, Brehm notes, compare the singing favorably to "that of a Canary or even of a Nightingale," whereas others are less enthusiastic, for instance a certain Herr Schacht, "a well-known educator and reliable and well-informed observer," who claims to have had a singing mouse for some time.

> Its song did not bear the slightest resemblance to the bright song of a Canary or the deep trills of a Nightingale. It was nothing but "a twittering, a mixture of long-drawn, squeaking, piping sounds," which in the quiet of night could be heard at a distance of twenty paces. The song of another Mouse, observed by Herr Mueller, another tutor, consisted of "soft, whistling sounds, uttered slowly or in a more lively manner, in the latter case reminding one distinctly of a bird's song, but being much weaker."
>
> (Brehm 1895, 338)

In any case, Brehm concludes, it would be "more congruous to speak of 'twittering' Mice than of 'singing' ones" (ibid.).

There are thus several layers of ambiguity surrounding the singing of mice: first, whether it should be characterized as singing at all, and, second, whether all mice sing or only certain exceptional individuals. What interests me about all of this is, first, the way it points to a world beyond human perception which science is only now beginning to be able to detect—that this is, in other words, an unheard, inhuman music—and, second, how this relates to the cultural imaginary surrounding mouse song, which hovers perennially at the boundary between sound and silence.

The most famous articulation of this imaginary is no doubt Franz Kafka's "Josefine, the Singer or the Mouse Folk," the last story he wrote and also the last to be published in his lifetime. Kafka was well-acquainted with Brehm's *Tierleben*, and so it is not unlikely that he was familiar with the tales of singing mice. Moreover, at the time he wrote "Josefine," Kafka was in the process of losing his voice to tuberculosis,[1] and the narrative follows a trajectory from sound to silence, ending with the disappearance of Josefine's voice. Like most of Kafka's late animal stories, it has a first-person narrator, who in this case is not Josefine but rather another mouse, presumably a male, who is describing the relationship between Josefine and the other mice. The entire narrative can be seen as an example of what Gerhard Neumann (1968) called Kafka's "gliding paradox," a peculiarly Kafkaesque rhetorical figure, whereby an initial affirmative statement—for example, "Our singer is called Josefine"—is gradually negated, whereupon that negation is itself negated, and so on, until all certainty has been eroded. In this story, the central tension concerns the nature and definition of Josefine's singing: having announced that Josefine is "our singer," the narrator then proceeds to question whether it really is singing, whether it isn't more of a whistling (*Pfeifen*), and then paragraph by paragraph the opening statement is questioned, qualified, and negated until there's nothing left: it's singing; no, it's more like whistling; but no actually it's not whistling; in fact it's less than the ordinary everyday whistling of all the other mice; her voice is really "nothing" (2007, 100); there is nothing musical about it, or if there is, "then it is reduced to the lowest possible nothingness" (102); no of course it is whistling, "[h]ow could it be anything else? Whistling is the language of our people" (103), and so on, until at last we learn that Josefine has disappeared, at which point the narrator asks whether there will be any noticeable difference between Josefine's absence and her presence, and whether the gatherings where she used to sing weren't in fact completely silent all along.[2]

The text as a whole resonates powerfully with questions concerning language, music, sound, and voices both human and nonhuman. Furthermore, because Josefine is not the narrator of her own story, she is in effect doubly silenced as we never actually "hear" her speak or sing in the text at all. The text thus revolves around not only the ambiguous distinction between singing and whistling, speaking and falling silent but ultimately also the question of narrative authority and in particular the category of the narrative *voice*. It is this ambiguously anthropomorphic category that I would like to explore in this chapter, specifically as it relates to zoopoetics and the question of what it would mean to hear the voice of an animal in or through a literary text.

2. Josefine Sings the Blues

Near the beginning of the second part of *The Animal That Therefore I Am*, having reeled off a long list of questions concerning what is "proper" to "the animal," Derrida wonders whether it would be possible to reimagine the discourse on the animal in musical terms, as a score or staff, and if so, whether one could change the key or the tone of the music by inserting a "flat" (♭)—a "blue note," if you will:

> I wish only to indicate a tonality, a pitch of the notes that change an entire staff [*une hauteur des notes qui changent toute une portée*]. How might one change the range [*la portée*] of such questions regarding the being of what would be proper to the animal? How might one add a flat, as it were, to the key of these questions and change the music? [*Comment, en quelque sorte, mettre un bémol à la clé de ces interrogations et changer la musique?*].
>
> (Derrida 2008, 63, translation modified)

The goal of this transposition or modulation, which he says would be "contradictory" or even "impossible," would be to "render audible" [*faire entendre*, also: "understandable"] "an unheard language or music" (*une langue ou une musique inouïe*) that would be "somewhat inhuman," but a language nonetheless, "not those inarticulate cries or insignificant noises, howling, barking, meowing, chirping, that so many humans attribute to the animal, a language whose words, concepts, singing, and accent can finally manage to be foreign enough to everything that, in all human languages, will have harbored so many asinanities (*bêtises*) concerning the so-called animal" (63, translation modified). The first *bêtise* to eradicate would thus be the one that reserves language and singing exclusively for human beings.

The idea of the "unheard"—*l'inouï*—is a recurring figure in Derrida's work, going all the way back to the inaudible distinction between the *e* and the *a* in "*différance*" (Derrida 1982, 22). It is thus, fundamentally, allied with the *trace*, with the movement of *différance* and differentiation, in short, the entire "network of possibilities [. . .] without which there would be no language" (Derrida 1995, 284–5). As a consequence of this, the "unheard" or "unheard-of" (both meanings of *inouï*) refers to an alterity or exteriority that cannot be assimilated to the text of philosophy, or which philosophy cannot domesticate and organize under the established rubrics of "the Other" and "the Outside," but which could in some way occupy a position outside, beyond, or prior to the classical dichotomies of self and other, man and animal, male and female, λόγος and φωνή, and so on, but also, importantly, beyond the very distinction between audible and inaudible, intelligible and unintelligible. The dream of rendering audible (*faire entendre*) this unheard language is "impossible" and "contradictory" precisely because, as Jean-Luc Nancy puts it in his book on *Listening*, the philosopher is "someone who always hears [*entend*] [. . .], but who cannot listen [*écouter*], or who, more precisely, neutralizes listening within himself, so that he can philosophize" (2007, 1). The "impossible" task thus involves not only finding a new mode of expression but also teaching philosophy to hear and understand (*entendre*) the question (of Being, of the animal, of language) *otherwise*: "that is," as Derrida puts it elsewhere, "within the openness of an unheard-of [*inouïe*] question that opens itself neither onto knowledge nor onto a non-knowledge as knowledge to come. In the openness of this question, *we no longer know*" (2011, 88, original emphasis). And it is within this space of openness and indeterminacy that an unheard, inhuman language or music might, perhaps, become audible.

This unheard language thus appears to be aligned with the unheard(-of) questions that deconstruction has always sought to open up, and, in this regard, we may observe an affinity between it and that which Derrida had, in the first part of his lecture, referred to as *la pensée de l'animal*—translated as "thinking concerning the animal" (2008, 7) but which could also equally refer to the thoughts *of* the animal itself; a thinking, in other words, that would not hinge on the a priori exclusion of "the animal." This animal thinking, he writes, "derives from [*revient à*, also: 'comes back to'] poetry. There you have a thesis: it is what philosophy has, essentially [*par essence*], had to deprive itself of. It is the difference between philosophical knowledge and poetic thinking [*une pensée poetique*]" (*ibid.*). Poetic thinking, by implication, is in some sense synonymous with "animal thinking," and both are *essentially* at odds with philosophical knowledge. Thus, if this unheard, inhuman language is to be made audible anywhere, it will be through literature, not philosophy.

It is not altogether surprising, then, that throughout this passage, Derrida appears secretly to be talking about Kafka. Just a few pages earlier he has reminded the audience that it was in this very *château* that he once spoke of "Freud and Kafka" (55), by which he means the Château de Cerisy-la-Salle, where the conference on the "autobiographical animal" was being held, but which also clearly recalls the title of Kafka's final novel, *Das Schloss* (*The Castle*; *Le Château*). This leads him to ponder the question of the subconscious and whether animals dream, which in turn leads him to his own dream of this unheard and inhuman music: "Before even beginning to dig into the burrow of words and images on the basis of which, in this château, I would dare address you, I dreamed for a long time" (62). And then, just after imagining how you could sneak a "blue note" into the musical score of the question of the animal, he repeats that he is "dreaming, therefore, in the depths of an undiscoverable burrow to come" (63). Clearly a reference to "The Burrow," in which the animal narrator likewise has an "impossible dream" of changing the entire structure of his burrow of words. Yet I wonder if the repeated references to music and singing aren't in fact a subterranean allusion to "Josefine," whose voice is thus literally "unheard" even in Derrida's text. The "dream," then, would be to make Josefine "sing the blues," as it were, in such a way that it would be meaningless to dispute whether this is singing or whistling but that she might nevertheless be heard.

3. Who Is Squeaking?

As mentioned, it is not Josefine who narrates her own story but rather an anonymous, intradiegetic narrator, who speaks on behalf of the "mouse folk" (*Volk der Mäuse*), and it is in the mode of a collective "we" that he questions Josefine's status as the voice of the people. In this sense, the text may be said to stage a conflict between the *figure* of the voice and the function of the *narrative* voice. Here it is certainly significant that the former, as embodied by Josefine, should be on the side of music, childishness, and femininity, all traditionally associated with the body, the sensuous, and the irrational, whereas the latter presents itself as implicitly masculine, paternalistic, rational, "unmusical" (2007, 95), and so on. The text itself appears to invite us to conceive of the relationship between Josefine and the community of mice in just this way, namely as that between a child (Josefine) and its father (the people) (99). Within this logocentric schema, the narrative voice would be the embodiment of λόγος, on the side of the Father and the Law, while Josefine, embodying φωνή, would be on the side of the animal, excluded from political life, "outside the law" (*außerhalb des Gesetzes*) (103). Yet this dichotomy is itself immediately undermined—"all of this is simply, absolutely, untrue" (103)—not only because the mice themselves are repeatedly characterized as childlike and because, in Kafka, there is nothing outside the law, but also, I would argue, because the narrative voice is itself also that of an *animal*. But can one even imagine a narrative voice that is anything other than human?

Certainly, within classical narratology, the concept of the "narrative voice" has always been implicitly human, and indeed humanist, invoking a presence and stable point of origin for the narrative, "a subjectivity intimately inhabiting the text" (Gibson 1996, 143), speaking to us. This is somewhat surprising, given how narratology has in almost all other respects sought to distance itself from hermeneutics and the humanist legacy. At the same time, it is perhaps understandable, at an intuitive level, given that humans, as far as we know, are the only animals that write (and read) literary texts. Even if we accept Roland Barthes's famous objection that "writing is the destruction of every voice, of every point of origin" (Barthes 1977, 142) and that therefore "in the text, only the reader speaks" (Barthes 1974, 151), even this voice, the voice which the reader lends to the text, will be a human voice. It is no doubt for this very reason that (literary) animal studies has for the most part ignored or avoided the term "narrative voice" when discussing the presence and agency of nonhuman animals in literary texts. And yet, the seemingly unassailable, always-already-human identity of the narrative voice causes problems on multiple levels—indeed, a salient characteristic of Kafka's animal narratives is how they exploit these problems to their advantage.

On the one hand, *voice* (φωνή) is neither synonymous nor coextensive with *speech* (λόγος). Hence, even if the latter has served as a privileged marker of the anthropological difference, the former is shared by all sensate beings and hence not reducible to the human—nor does every human vocal utterance constitute speech. On the other hand, the status of language as a distinguishing feature of the human is itself highly problematic and ultimately untenable, unless, of course, one consents to define "language" in such a way that it automatically excludes "the animal." But there is essentially no good reason to accept such a definition, since, as Derrida, for example, insists, the "network of possibilities" that make language possible in the first place—that is, the trace, iterability, *différance*, etc.—"*are themselves not only human*" and hence do not "give rise to a single linear, indivisible, oppositional limit" between the human and the nonhuman (1995, 284–5, original emphasis). So-called "human" language would thus need to be considered as fundamentally inseparable from other "forms of marking," including "the complexity of 'animal languages,' genetic coding," and so forth (285). Indeed, as posthumanists like Cary Wolfe never tire of emphasizing, what we tend to call human language is itself a form of prosthetic "technicity or mechanicity," meaning that

> "we" are always radically other, already in- or ahuman in our very being—not just in the evolutionary, biological, and zoological fact of

> our physical vulnerability and mortality, which we share, as animals, with animals, but also in our subjection to and constitution in the materiality and technicity of a language that is always on the scene before we are, as a radically ahuman precondition for our subjectivity, for what makes us human.
>
> (Wolfe 2009, 571)

In other words, "speech" is not reducible to the human any more than "voice" is; "speech" is not and never has been constitutively or exclusively "human" at all.

What does this mean for our conception of narrative and of the narrative voice? Perhaps we should begin by asking what we mean by "narrative," a term which is by now so ubiquitous as to have lost all specificity. In an effort to counteract this diffusion, in his book *Modernism, Narrative and Humanism*, Paul Sheehan proposes the following basic definition: "Narrative, the process of storymaking and storytelling, is language arranged meaningfully over time" (2004, 9). The relationship between these three key elements—language, meaning, and time—is complex, and it is this complexity that constitutes the narrative. In narrative, disparate events are joined together to form a series through a logic of causality and mutual implication—one thing leads to another—and through this process, otherwise known as the "plot," the events narrated become imbued with meaning. Referring to Frank Kermode's definition of plot as "an organization that humanizes time by giving it form" (*ibid.*), Sheehan concludes that narrative "is *human-shaped*. It is a uniquely human way of making order and meaning out of the raw material of existence. . . . Put simply, we tell stories about ourselves to give our lives meaning and purpose, and about our kind to maintain the crucial human/inhuman distinction" (ibid., original emphasis). The human, in short, is a storytelling species—an "autobiographical animal," to coin a phrase—whose self-image and identity are constituted in and through narrative practices. This applies to myths and folktales just as it does to the discourse of Western, humanist anthropocentrism. In other words, the "human shape" that Kermode attributes to narrative as such is itself a product of particular narratives that have emphasized certain aspects while suppressing others in order to produce a sense of necessity and order. Narrative, then, is itself a version of the "anthropological machine" (Agamben 2004), a mechanism for producing the recognition of the human, via the "inclusive exclusion" of the animal.[3] In this case, narrative is a machine for transforming contingency into necessity, chaos into meaning.

And yet, as Sheehan goes on to observe, in order for narrative to appear "human-shaped," there has to be something in it that pushes against the suppression of contingency and difference that lies at the heart of narrative logic. Otherwise, all stories would be the same. In order to account for this, Sheehan reintroduces the concept of voice, proposing a model of literary narrative as "a composite of voice and machine," where "machine" refers to the mechanics of plot and causality, while "voice" refers to "difference, variation and irregularity" (11). Narrative would thus be characterized by an oscillation between necessity (machine) and contingency (voice). The more "successful" a narrative is in eliminating randomness and contingency—that is, the more perfectly it conforms to expectations and generic conventions—the less "human" it begins to seem. After all, we tend to prefer stories that are not entirely predictable, that seem to depart from the norm in unexpected and innovative ways. On the other hand, if "voice" takes over, then the sense is lost. The "voice-machine complex," Sheehan writes, can also be thought of in terms of the "play between 'difference' and 'sameness'" (174). It is important to note that at *both* ends of this spectrum represent the end of meaning and signification: either through pure automatism and the "smooth functioning" of mechanicity which eliminates all difference (roughly equivalent to what Roland Barthes calls the "rustle" of language)[4] or through the total contingency and randomness of pure vocality, which would eliminate all sameness. Only in the interplay of these two extremes can narrative and meaning emerge.

The narrative innovations of modernist literature, Sheehan's principal object of study, can thus be read in terms of experimentation with the "voice-machine complex." And indeed, in order to see this complex at work, one need only open a page of Kafka, where both machines and voices play a crucial role, invariably posing a threat to narrative order, coherence, and meaning. Sheehan's conception of narrative can thus help us to account for the internal tensions and discontinuities at work in Kafka's writings. Nevertheless, I find it surprising that in his effort to configure both "human and nonhuman claims" (10) about language and narrative, Sheehan consistently equates "voice" with the human and "machine" with the nonhuman. Why would the contingency and disruption that he identifies with voice be uniquely human? Or, to put it another way: why would Sheehan, whose entire project runs counter to Cartesian humanism, place the animal on the side of the machine? In so doing, he runs the risk of affirming precisely the assumption that he and the modernist authors he reads call into question, namely the constitutive anthropomorphism of narrative as such. "Because narrative is voice *and* machine," Sheehan writes, "its human countenance is complicated by a nonhuman infrastructure" (175, original emphasis). No doubt, but this supposedly human countenance is *also* complicated by the constitutively a-, in-, or more-than-human nature of the voice itself. In other words, much like Agamben's anthropological machine, Sheehan's voice-machine complex presents as a binary what is in fact better understood as what Dominic Pettman calls a "cybernetic triangle," that is, an "unholy trinity of human, animal, and machine" (2011, 5). This would go some way toward liberating the voice from its unquestioned association with human(ist) agency and subjectivity.

With this in mind, let us now revisit Sheehan's minimal definition of narrative as "language arranged meaningfully over time" and compare it to the definition of "song" employed in the aforementioned studies of ultrasonic mouse vocalizations: "a sound of animal origin that is not both accidental and meaningless," consisting of "a series of notes or syllables, generally of more than one type, uttered in succession and so related as to form a recognizable sequence or pattern in time" (Holy and Guo 2005, 2178). The resemblance is quite striking. In fact, the two definitions are practically identical, to the extent that, *mutatis mutandis*, the latter could plausibly serve as a definition of human narrative, particularly if we clarify *to whom* this purposeful (non-accidental) utterance is supposed to be meaningful (or at least not meaningless). Moreover, the rather awkward negation/conjunction "not both . . . and," with its implicit distinction between intentionality and interpretation, is basically a hermeneutic theory in miniature. Conversely, if we apply Sheehan's definition to the songs of mice, the question arises: can we identify a "voice-machine complex" there as well? Arguably, this is precisely what the 2015 study mentioned at the outset (Chabout et al. 2015) revealed, namely that the songs of mice are not, in fact, simply mechanical and repetitive but rather vary according to complex social cues and environmental circumstances. Moreover, these ultrasonic vocalizations seem to comprise both innate and learned elements (cf. Arriaga and Jarvis 2013), meaning that the individual singer is able to introduce variations and permutations into the mouse song "canon."

Now, to be clear, I am not claiming that there is *no difference* between an ultrasonic mouse vocalization and a literary text written by a specific human being, for example, Franz Kafka. Nor do I wish to imply that these diverse practices exist on a scale or within a hierarchy that would once again place human modes of *poeisis* and narrative "above" those of other species. Rather, I am proposing merely that the structural similarity of these two definitions should serve as a further reminder, to quote Derrida again, that there is no "single, linear, indivisible oppositional limit" between the human and the nonhuman to be drawn on the basis of language and that the growing scientific knowledge surrounding "the complexity of 'animal languages' . . . does not allow us to 'cut' once and for all where we in general would like to cut" (1995, 285). Negotiating this zone of indeterminacy, while still paying careful attention to the specifically literary character of particular texts, and indeed, to quote Susan McHugh, how animals and animality function in those texts "as a function of what we think of as their literariness" (2011, 7)—that, I would say, is precisely the task of zoopoetics as I conceive of it, namely as both a mode of writing and a method of reading. In other words, while it is crucial to acknowledge that human and nonhuman modes of communication, including speaking and singing, are evolutionarily related processes, existing on a continuum and not as a strict binary, this does not ultimately tell us very much about the specific ways these processes are at work in any given literary text. Kafka's "Josefine" is not just a story about mice: it is also, importantly, a story about *art*, about language and music; it is a literary text that is fundamentally about its own status as literature. Moreover, this self-reflexivity is inextricably bound up with its engagement with the question of the animal. This is what makes it a zoopoetic text.

This is important to keep in mind, especially because Kafka chooses to approach the essentially poetic problem of literature and writing in terms of music, and, indeed, an unheard, inhuman music. As Burkhard Müller observes, in order to grasp the core of language, Kafka must seemingly "transpose it to a place where it loses its potential to communicate and is reduced to mere sound" (2010, 113). The text is thus predicated on a double transposition, first from the world of humans to the world of mice and, second, from speech to song. This double transposition is then followed by a third, namely from sound to silence, and this silence, crucially, is a silence of *writing*, a silence in words. Strictly speaking, there is no voice, either human or nonhuman, *in* the text. What the text does is take this absence of voice and transform it into a (zoo)poetic principle. Hence, although music is a central motif in his late animal stories, "Kafka is at pains to prevent it from entering the realm of audibility":

> The language Kafka employs attests to the overpowering force of music without giving a hint of what it would sound like—and does so by assuming that quality itself. It plays on syntax like a musician plays on keys and strings; their mechanical properties, however, tell you nothing of the notes that stream forth.
>
> (Müller 2010, 113–14)

Although he does not employ this idiom, what Müller is describing here is ultimately quite close to what Deleuze and Guattari mean by a "minor literature" which deterritorializes language and puts it to "strange and minor uses" (1986, 17). As Deleuze writes in a late essay, authors like Kafka "invent a minor use of the major language within which they express themselves entirely; they minorize this language, much as in music, where the minor mode refers to dynamic combinations in perpetual disequilibrium" (1997, 109). Language thus modulated into a minor key (or mode) is always at the limit and on the verge of breaking down. It "stutters," as Deleuze puts it; it introduces a disturbance into the smooth functioning of language by tapping into

> a line of variation or subtended modulation that brings language to this limit. And just as the new language is not external to the initial language, the asyntactic limit is not external to language as a whole: it is the outside of language, but is not outside it. It is a painting or a

> piece of music, but a music of words, a painting with words, a silence in words, as if the words could now discharge their content.
> (Deleuze 1997, 112–13)

One of the principal ways that Kafka achieves this "stuttering" is through the asignifying sounds and unheard, inhuman music that pervade especially his animal texts. In Deleuzian terms, the series of transpositions Müller describes could be conceived as successive stages of deterritorialization. First, speech is stripped of signification, "reduced to mere sound," whereupon that sound is itself reduced to silence, but this silence is a silence *in* and *of* the text and of language itself: "a silence in words."

4. Mouse, Interrupted

In what is to my mind still one of the most compelling readings of the story, Margot Norris draws attention to the rather surprising punctuation of the title: "Josefine, the Singer or the Mouse Folk." Ordinarily, you would read the "or" as separating the two titles,[5] but in view of the ambiguous relationship between Josefine and the rest of the mice as thematized in the text, there is also the intriguing possibility of reading "Josefine" as not only "the Singer" but also "the People" (*das Volk*). Or, to quote Norris: "the opposition is not between Josefine and the mouse folk but between Josefine's identity as a singer and her membership in the pack. The story poses the conundrum, *Is Josefine singular or is she plural?*" (120, original emphasis). Is hers an individual voice, or is it the voice of the people? In keeping with the general subversion of binary oppositions, the space of indeterminacy opened up by the "unheard" language or music, it seems that the only possible answer is that we do not know and that she can and must be regarded as both singular *and* plural. In fact we might go so far as to say that this is the irresoluble tension that drives the entire text.

There is one point in the narrative where this tension comes close to resolution, namely when the narrator describes the dreams of the mice—which, in fact, come quite close to Derrida's dream of making Josefine's singing audible by rendering the distinction between singing and whistling inoperative. In the "scant pauses between battles," the narrator says, the mice dream of being united, "stretch[ed] out in the big, warm, communal bed. And here and there into these dreams comes the sound of Josefine's whistling; she calls it sparkling (*perlend*), we call it stuttering (*stoßend*); but whatever it is, this is where it belongs more than anywhere else, in the way that music hardly ever finds the moment that is waiting for it" (Kafka 2007, 102–3, trans. mod.). This passage is remarkable, since the narrator effectively (albeit only momentarily) concedes that this is music, no matter what you call it, because of the *effect* it has, namely that of establishing a community, where everyone may find consolation in their neighbor's fur (100). "Naturally, it is a whistling," he says, "how could it be anything else? Whistling is the language of our people." They all whistle, but Josefine's whistling is the voice that emerges, impossibly, out of this collective.

The nature of Josefine's singing and her relation to the collective are not the only paradoxes of the story. At the end, the narrator says that Josefine will be forgotten, because "we [the mice] practice no history [*da wir keine Geschichte treiben*]" (108), but this too is a paradox, of course, since he is indeed telling a story (*Geschichte*), even if it is a self-effacing one. And hence the foundation of the narrative, and by extension the community of mice, cannot be seen simply as "silence" or as the "absence" of voice and history, but rather history *under erasure*, preserved in its negation: a "no-history." By the same token, at the beginning of the story, the narrator informs us that even though the mice are thoroughly unmusical, they nevertheless have an "inkling" (*Ahnung*) of what song is, and that some ancient songs have been preserved, although, of course, "no one can sing them anymore" (Kafka 2007, 95). But in any case, Josefine's singing bears no resemblance to these songs of legend. The entire narrative is thus framed in terms of a forgotten or erased communal history, which is nevertheless preserved as an "inkling" that can only be described in negative terms: it is *not* like Josefine's singing but *no one* can say what it *is* like.

In conclusion, then, I would like to suggest—all too briefly—that Josefine's ambiguously singular-plural identity and her status as the singer of her community resonate quite powerfully with Jean-Luc Nancy's discussion of founding myths (or legends) in *The Inoperative Community*. All community, he writes, is founded on myth, and all myth is "the myth of community" (1991, 51), which is to say that it carries with it a sense of completion or fulfillment. This is true both of ancient myths and of folktales as well as the more modern myths or "grand narratives" of humanist anthropocentrism: "Myth, in short, is the transcendental autofiguration of nature and of humanity, or more exactly the autofiguration—or the autoimagination—of nature as humanity and of humanity as nature" (54). In modernity, this form of immanent, totalizing community has become untenable, co-opted by fascist totalitarianism. The only alternative, however, appears to be liberal humanist individualism, which is synonymous with capitalist exploitation and the myth of progress. Thus, the only hope, as Nancy sees it, is to "interrupt" the myth of community, but to do so without negating it entirely, that is, without turning it into complete silence, say, since this too can be appropriated and put to work for the dialectical progression toward totality. The term he gives to this interruption that forestalls the putting to work of myth is "inoperativity" (*désœuvrement*): not total silence but rather the absence or the interruption of sound.

> In the interruption of myth something makes itself heard, namely, what remains of myth when it is interrupted—and which is nothing if not the very voice of interruption, if we can say this.

> This voice is the *voice of community*, . . . of the interrupted community, the voice of the incomplete, exposed community speaking as myth without being in any respect mythic speech. . . . When a voice, or music, is suddenly interrupted, one hears just at that instant something else, a mixture of various silences and noises that had been covered over by the sound, but in this something else one hears again the voice or the music that has become in a way *the voice or the music of its own interruption*: a kind of echo, but one that does not repeat that of which it is the reverberation.
>
> (62, emphasis added)

"A name has been given to this voice of interruption," Nancy continues, and that name is "literature" (63). By this, Nancy does not mean literary texts or anything that is clearly specific to human cultures and practices. Rather, the point is that the "literary" as such, its singularity, *is* the interruption of the myth of community. Nothing can follow from it: "this inaugural act founds nothing, entails no establishing, governs no exchange; no history of community is engendered by it" (68). In other words, through the "singular eruption" (ibid.) of this "nothing" of a voice, "we" practice "no history." Hence this voice does not found a new myth or a new homogeneous community but is always at the limit (67), and this limit can and must also be read as the "abyssal limit" (Derrida 2008, 12) between "man" and "animal."

Hence, perhaps we can risk positing that in order for literature to be the "voice" of a community that does not exclude "the animal," in order for this inoperative community to call into question precisely that distinction (without, however, claiming to have overcome or abolished it, which would, again, plunge us into the realm of mythology), this literature must be *zoopoetic*. Perhaps this is what Kafka's story reveals: Josefine's voice is the silent voice of an inoperative, inhuman community, singing the music of its own interruption. And what is this other than the unheard, inhuman music Derrida was dreaming of? Perhaps, then, it is not a question of inserting a *flat* into the score, but rather a *rest* (𝄽). And if we listen closely, if we learn to listen "otherwise," we may be able to hear that the rest is not silence but the interruption of song.

Notes

An earlier version of this chapter was originally published as "An Unheard, Inhuman Music: Narrative Voice and the Question of the Animal in Kafka's 'Josephine, the Singer or the Mouse Folk'," *Humanities* 6.2. (2017).

1. Upon completing the story, Kafka is reported to have remarked to his friend Robert Klopstock that he had begun his "investigation into animal squeaking" (*Untersuchung des tierischen Piepsens*) (Kafka 1977, 495n20) at just the right time.
2. The constitutive uncertainty surrounding Josefine's vocal performances is compounded in translation, specifically with regard to the narrator's use of the

word *Pfeifen*. The Muir translation, until relatively recently the only English version of the text, renders it somewhat awkwardly as "piping" (Kafka 1971, 361), whereas more recent translations give it as "whistling" (Kafka 2015, 229, 2016, 103). Stanley Corngold, however, opts for "squeaking," explaining in a particularly revealing footnote that "the German word translated here as 'squeaking' is *pfeifen*, which, *for human beings*, means 'whistling'" (Kafka 2007, 95n1, emphasis added). He thus gratuitously enforces the human-animal binary in a way that I would argue is at odds with the story itself. In most other respects, however, Corngold's translation is the closest to Kafka's text, and so in what follows I will refer to that edition, silently substituting "whistling" for "squeaking" throughout.

3. Let us remember that the inclusive exclusion of the animal within the human, of ζωή within βίος, which is the founding gesture of politics, is structurally identical to the inclusive exclusion of φωνή within λόγος (cf. Agamben 1998, 7–8). By the same token, Josefine cannot be said to be "outside the law"—at most she is "exclusively included" within it, as an exception.
4. "The rustle," for Barthes, "is the noise of what is working well [*Le bruissement, c'est le bruit de ce qui marche bien*]. From which follows this paradox: the rustle denotes a limit-noise, an impossible noise, the noise of what, functioning to perfection, has no noise [*pas de bruit*]; to rustle [*bruire*] is to make audible [*faire entendre*] the very evaporation of noise" (Barthes 1986, 76–77). The "rustle" of language is the sound of the limit, the impossible noise of the linguistic machine functioning to perfection, which means that we do not hear it. It is, hence, "the noise of an absence of noise" (ibid., 78), the sound of silence. Consequently, it is also impossible to determine whether this sound is a singing or a whistling. It is in this sense that we may begin to understand the murine narrator's verdict that Josefine's voice is a "nothing" (*dieses Nichts an Stimme*), but a nothing which nevertheless asserts itself and has an effect (*Wirkung*) on the mice, who, it turns out, "really [*wirklich*] do listen" to her (Kafka 2007, 100–101, trans. mod.).
5. And, indeed, with this exception of Corngold's, all the aforementioned English translations "correct" or normalize this unorthodox punctuation.

References

Agamben, Giorgio (1998). *Homo Sacer: Sovereign Power and Bare Life*. Translated by Daniel Heller-Roazen. Stanford: Stanford University Press.

Agamben, Giorgio (2004). *The Open: Man and Animal*. Translated by Kevin Attell. Stanford: Stanford University Press.

Arriaga, Gustavo, and Jarvis, Erich D. (2013). "Mouse Vocal Communication System: Are Ultrasounds Learned or Innate?" *Brain and Language*, no. 124: 96–116.

Barthes, Roland (1974). *S/Z*. Translated by Richard Miller. New York: Hill & Wang.

Barthes, Roland (1977). "The Death of the Author." In *Image-Music-Text*, translated by Stephen Heath, 142–8. New York: Hill & Wang.

Barthes, Roland (1986). "The Rustle of Language." In *The Rustle of Language*, translated by Richard Howard, 76–9. New York: Farrar, Straus and Giroux.

Brehm, Alfred Edmund (1895). *Brehm's Life of Animals: A Complete Natural History for Popular Home Instruction and for the Use of Schools, Vol. 1: Mammalia*. Translated from the Third German Edition as edited by Pechuel-Loesche and William Haacke, and revised and abridged by Richard Schmidtlein. Chicago: A. N. Marquis.

Chabout, Jonathan, Sarkar, Abhra, Dunson, David B., and Jarvis, Erich D. (2015). "Male Mice Song Syntax Depends on Social Contexts and Influences Female Preferences." *Frontiers in Behavioral Neuroscience* 9: 76.

Deleuze, Gilles (1997). "He Stuttered." In *Essays Critical and Clinical*, translated by Daniel W. Smith and Michael A. Greco, 107–14. Minneapolis: University of Minnesota Press.

Deleuze, Gilles, and Guattari, Félix (1986). *Kafka: Toward a Minor Literature.* Translated by Dana Polan. Minneapolis: University of Minnesota Press.

Derrida, Jacques (1982). "Différance." In *Margins of Philosophy*, translated, with additional notes by Alan Bass, 1–27. Chicago: University of Chicago Press.

Derrida, Jacques (1995). "'Eating Well', or the Calculation of the Subject." Translated by Peter Connor and Avital Ronell. In *Points . . . Interviews, 1974–1994*, edited by Elisabeth Weber, 255–87. Stanford: Stanford University Press.

Derrida, Jacques (2008). *The Animal That Therefore I Am.* Edited by Marie-Louise Mallet. Translated by David Wills. New York: Fordham University Press.

Derrida, Jacques (2011). *Voice and Phenomenon: Introduction to the Problem of the Sign in Husserl's Phenomenology*. Translated by Leonard Lawlor. Evanston: University of Illinois Press.

Dunne, Rob (2011). "The Mystery of the Singing Mice." *Smithsonian Magazine*, May 11.

Gibson, Andrew (1996). *Towards a Postmodern Theory of Narrative*. Edinburgh: Edinburgh University Press.

Holy, Timothy E., and Guo, Zhongsheng (2005). "Ultrasonic Songs of Male Mice." *PLoS Biology*, no. 3: 2177–86.

Kafka, Franz (1971). "Josephine the Singer, or the Mouse Folk." Translated by Willa and Edwin Muir. In *The Complete Stories*, edited by Nahum N. Glatzer, 360–76. New York: Schocken Books.

Kafka, Franz (1977). *Letters to Friends, Family, and Editors*. Translated by Richard and Clara Winston. New York: Schocken Books.

Kafka, Franz (2007). "Josefine, the Singer or the Mouse People." In *Kafka's Selected Stories*, translated and edited by Stanley Corngold, 94–108. New York: W.W. Norton.

Kafka, Franz (2015). "Josefine, the Singer, or the Mouse People." In *Metamorphosis and Other Stories*, translated by Michael Hofmann, 228–44. London: Penguin Classics.

Kafka, Franz (2016). "Josephine, Our Meistersinger, or the Music of Mice." In *Konundrum: Selected Prose of Franz Kafka*, translated by Peter Wortsman, 102–28. New York: Archipelago Books.

McHugh, Susan (2011). *Animal Stories: Narrating Across Species Lines*. Minneapolis: University of Minnesota Press.

Müller, Burkhard (2010). "Consolation in Your Neighbor's Fur: On Kafka's Animal Parables." In *Kafka's Creatures: Animals, Hybrids, and Other Fantastic Beings*, edited by Marc Lucht and Donna Yarri, 119–35. Lanham, MD: Lexington Books.

Nancy, Jean-Luc (1991). *The Inoperative Community*. Edited by Peter Connor. Translated by Peter Connor, Lisa Garbus, Michael Holland, and Simona Sawhney. Minneapolis: University of Minnesota Press.

Nancy, Jean-Luc (2007). *Listening*. Translated by Charlotte Mandell. New York: Fordham University Press.
Neumann, Gerhard (1968). "Umkehrung und Ablenkung: Franz Kafkas 'Gleitendes Paradox'." *DVjs*, no. 42: 702–44.
Norris, Margot (1985). *Beasts of the Modern Imagination: Darwin, Nietzsche, Kafka, Ernst, & Lawrence*. Baltimore: Johns Hopkins University Press.
Pettman, Dominic (2011). *Human Error: Species-Being and Media Machines*. Minneapolis: University of Minnesota Press.
Sheehan, Paul (2004). *Modernism, Narrative, and Humanism*. Cambridge: Cambridge University Press.
Wolfe, Cary (2009). "Human, All too Human: 'Animal Studies' and the Humanities." *PMLA* 124, no. 2: 564–75.

13 From Being a Beast to Becoming-Animal

Literary Experiments in Crossing the Species Divide

Jonathan Pollock

This chapter proposes to re-examine the endless debate about the impossibility of avoiding anthropocentrism and anthropomorphism when writing about animals. Here is a recent definition of these two pitfalls:

> The anthropocentrists describe the natural world as it appears to humans. Since they are writing books for humans, that's perhaps commercially shrewd. But it is rather dull. The anthropomorphists assume that animals are like humans: they dress them in actual . . . or metaphorical . . . clothes and give them human sense receptors and cognition.
>
> (Foster, xii)

Charles Foster, the author of these lines, is an Oxford don and veterinary surgeon who set himself the task of "Being a beast," or rather of being five beasts, and whose account of these attempts was published in 2016.

> I describe the landscape as perceived by a badger, a fox, an otter, a red deer and a common swift. I use two methods. First, I immerse myself in the relevant physiological literature and discover what is known from the laboratory about the way these animals function. Second, I immerse myself in their world. When I'm being a badger, I live in a hole and eat earthworms. When I'm being an otter, I try to catch fish with my teeth.
>
> (Foster, xii)

In other words, he does not simply observe animals in their natural environment; he adopts their habitat and life-style. When *being* a badger, he lives in a sett on the Welsh hillside and shuffles along on all fours. When being an otter, he dons a wetsuit and swimming goggles, glides down moorland streams, and leaves piles of excrement, or spraints, to mark his territory. As an urban fox, he rummages through dustbins and takes 40 winks in backyards. As a red deer, he experiences what it is like to be hunted by asking a friend to set a bloodhound on his tracks. Finally, to be a swift—the most challenging aspiration of all—he takes up hang-gliding, climbs trees in order to gulp down airborne insects, and travels in the wake of migrating swifts from England to West Africa. Needless to say, Charles Foster is British: this is nature writing at its most eccentric.

Taking his inspiration from Jakob von Uexküll, Foster conducts forays into the multiplicity of animal worlds:

> When we walk into a wood, we share its sensory outputs (light, colour, smell, sound and so on) with all the other creatures there. But would any of them recognise our description of the wood? Every organism creates a different world in its brain. It lives in that world. We are surrounded by millions of different worlds. Exploring them is a thrilling neuroscientific and literary challenge.
>
> (Foster, xi)

However, and despite some astonishing insights, Charles Foster is too much of a scientist to believe that he can succeed in perceiving the environment from anything but a human perspective:

> I'd hoped to write a book that had little or nothing of me in it. The hope was naïve. It has turned out to be (too much) a book about my own rewilding, my own acknowledgement of my previously unrecognised wildness, and my own lament at the loss of my wildness.
>
> (Foster, xv)

There is nothing more exhilarating than failure, and this is a truly exhilarating book, but I can't help feeling that Charles Foster is a victim of his own scientism. There is no doubting his physical courage and stamina, but he stops short of what he himself calls, in a throw-away phrase, "literary shamanism" (Foster, xi). Literature and shamanism: two ways of leading us out of our own worlds and minds and into those of others, two ways not of *being a beast* but, to quote the title of a chapter from *A Thousand Plateaus* by Gilles Deleuze and Félix Guattari, of "Becoming-Intense, Becoming-Animal, Becoming-Imperceptible." If you can never *be* a beast, it might be possible to *become* one, providing that the process of becoming never reaches its term or end.

From an anthropological point of view—and to use Philippe Descola's fourfold categorisation of cultural ontologies in *Beyond Nature and Culture*—Charles Foster remains a naturalist and eschews animism. To his mind, animism (or shamanism) is more or less a disguised form of anthropomorphism. This is why he takes umbrage with what he calls the "quasi-shamanism" of John Alec Baker, whose master-piece, *The Peregrine*, first published in 1967, might be thought to have done for that species what Foster is trying to do for five others. If the accuracy of Baker's natural history has sometimes been called into doubt, there is no denying his literary genius. "I have always longed to be part of the outward life," Baker writes, "to be out there at the edge of things, to let the human taint wash away in emptiness and silence as the fox sloughs his smell into the cold unworldliness of water; to return to the town as a stranger" (Baker, 28). To assuage this longing, Baker decides to pursue a peregrine falcon:

> Wherever he goes, this winter, I will follow him. I will share the fear, and the exaltation, and the boredom, of the hunting life. I will follow him till my predatory human shape no longer darkens in terror the shaken kaleidoscope of colour that stains the deep fovea of his brilliant eye. My pagan head shall sink into the winter land, and there be purified.
>
> (Baker, 48)

Over the months (which were actually years), Baker draws steadily closer to the hunter he hunts. On one occasion, he comes across the butchered body of a woodpigeon:

> I found myself crouching over the kill, like a mantling hawk. My eyes turned quickly about, alert for the walking heads of men. Unconsciously I was imitating the movements of a hawk, as in some primitive ritual; the hunter becoming the thing he hunts.

Accordingly, the pronouns shift from "I" to "we":

> We live, in these days in the open, the same ecstatic fearful life. We shun men. We hate their suddenly uplifted arms, the insanity of their flailing gestures, their erratic scissoring gait, their aimless stumbling ways, the tombstone whiteness of their faces.
>
> (Baker, 92)

He describes men not as they appear to other men, but as they might appear to a peregrine falcon, for whom other life-forms are defined less by their anatomical shapes than by the ways they move and the gestures they make. Here, the evaluation of characteristic human gestures ("impression") is coloured by an immediate emotional response ("judgement"). To the eyes of a creature whose own bodily movements are so much more rapid and finely tuned than those of any mammal, human gestures can only seem clumsy, inefficient, and hateful.

Now, it is precisely this pretence of becoming a falcon that Foster objects to:

> If J.A. Baker really disappears, who is left to tell the story? And if he doesn't, why should we take the story seriously? . . . Always we're left with the question: Who's speaking here? A peregrine given a

> Cambridge education? Or Baker peregrinised? Because we're never quite sure, the method never quite convinces. It's of the nature of poetry that it never quite declares its hand.
>
> (Foster, 6)

Foster, for his part, prefers to be as honest as possible:

> It might be rather prosaic to be able to say of every passage in the book, "This is Charles Foster writing about an animal", rather than "This might be a mystical utterance from a man-badger," but it's a lot less confusing.
>
> (Foster, 6–7)

Poetic prose is to be suspected. Mundane prose is more reliable.

I would like to suggest that Foster's rejection of "literary shamanism" is based on at least three presuppositions. The first concerns his understanding of Uexküll's animal worlds, the second his conception of becoming, the third his attitude towards the practice of writing.

It is difficult to overestimate the enormous influence Uexküll's writings have had on modern thought, an influence that extends far beyond the realms of ethology and biology. The philosophical concepts of "subject" and "image" have been considerably altered by his findings. The subject is no longer understood as standing in opposition to, or in correlation with, an object but as immersed in a milieu, a world (*Umwelt*). The milieu is circumscribed by the sensorial, affective, and motor capacities of the subject, who may thus be defined as one of any number of perspectives upon one of any number of worlds. Broadly speaking, the idea that every animal lives in a perceptually delimited environment peculiar to that animal has been understood in one of two ways. The first, epitomised by Martin Heidegger in his lectures from 1929–1930, and to which Maurice Merleau-Ponty pays lip-service in his course notes on nature (1957–1958), assumes that (1) "the *Umwelt* is a closure which cuts the animal off from the vast majority of external stimuli" (Merleau-Ponty, 224); and (2) the *Umwelt* is particular to a given species, so that, unless you belong to that species, it is perfectly inaccessible. And, admittedly, there is evidence to suggest that Uexküll thought of animal worlds as being like hermetically sealed monads, although he does also envisage their being interrelated via some sort of pre-established harmony or symphony of nature. But, even more interestingly, he stresses the metamorphic nature of the diverse *Umwelten*:

> With the number of actions available to an animal, the number of objects in its environment also increases. It increases as well in the individual life of any animal capable of accumulating experiences, for each new experience conditions a new attitude toward new impressions.
>
> (Uexküll, 96)

If the capacities of an individual animal evolve over the course of its life, so do those of its species over the course of myriad generations. Therefore, from both an ontogenetic and a phylogenetic perspective, *Umwelten* do not remain rigidly identical: they also become . . .

Foster's experience seems to corroborate Heidegger's reading of Uexküll. Does this mean that *Umwelten* cannot communicate? Are animals unable to pass from one world to another? Are they not *open* to the *Umwelten* of other species? Merleau-Ponty points out that if the *Umwelt* is the result of a projection, if it does represent each organism's way of making sense of its surroundings, animals also produce forms which are visible for others, whoever or whatever they may be: "Animals (Portmann): the body as an organ for others (*pour-autrui*)" (Merleau-Ponty, 288). For Alfred Portmann, an animal is a form of expression and communication from the very moment its body offers itself up to the interpretative gaze of another subject. Inasmuch as they are "organs to be seen," the form and gestures of an animal may be construed as being "addressed," and therefore of "expressing" something to somebody else: they reveal a kind of intersubjectivity already at work in the morphology of even those organisms which possess neither sight-organs nor the capacity to apprehend things as objects.

It is precisely this intuition which led Merleau-Ponty to believe that conventional articulated human language could only have emerged from a bedrock of natural symbolism ("symbolisme d'indivision," prior to the separation of signifier and signified). "Each one of us, said [Paul] Valery is an 'animal of words.' Reciprocally, you could say that animals (*animalité*) are the *Logos* of the sensory world" (Merleau-Ponty, 219). According to Portmann, the expressivity of their shapes and colours and movements extends far beyond the functional semiotics of intra-specific communication. But is this proof enough that we can enter their worlds and adopt their point of view—inasmuch as the point of view defines the subject, and not the subject the point of view? As we have seen, Charles Foster concludes from his own experience that it is humanly impossible to adopt the perspective of an animal. "Being a beast" enables him to reawaken the sensory capacities of his own body, while failing to attain the perceptual and motor abilities of the animals he imitates. As he himself points out, inter-species perspectivism is really only possible from within an animist mindset or *Weltanschauung*, a possibility that he refuses to entertain for himself and for those of a similar background (Baker). Fortunately there exist at our disposal a number of ethnographical and anthropological studies written by authors whose field-experience consists in years of communal living with tribal groups in the Amazonian Rain Forest. In the wake of Claude Lévy-Strauss, researchers such as Philippe Descola, Eduardo Viveiros de Castro, Eduardo Kohn, and Tim Ingold have learnt to take animist thought seriously, patiently translating its complexity into a conceptual language that makes it (partly) accessible to non-animist readers. Kohn, in *How Forests Think* (2013), and Viveiros de Castro, in *Cannibal Metaphysics* (2009), both devote long developments to the role of the shaman in inter-species communication:

> Native American shamanism may be defined as the ability possessed by certain individuals to cross the inter-species divide and to adopt the perspective of subjectivities from other species, in view to administer relations between the latter and human beings. By seeing non-human beings in the same way as these non-human beings see themselves (as humans), shamans are able to take on the role of active participants in a dialogue between species; and above all they are able to return [to their human perspective] in order to relate their story, something that the uninitiated cannot do. The meeting or exchange of perspectives is a dangerous process, and a political art—a form of diplomacy. If Western relativism has multiculturalism for public policy, Native American shamanistic perspectivism has multi naturalism as cosmic policy.
>
> (Viveiros de Castro, 25)

The notion of multinaturalism is an explicit reference to the plurality of animal worlds explored by Uexküll.

The second of Charles Foster's presuppositions concerns the notion of becoming. Despite his interest in, and love of, animals, Foster is closer to Descartes than to Aristotle. You are a hawk or you aren't. The choice is between fully fledged actuality or nothingness; there is no in-between state, no dimension of potential reality. However, if you are on your way to becoming something, then you haven't yet become it. What is more, if you subtract from the reality of becoming all notion of an end, a *telos*, a definitive form or entelechy, then you are left with a vision of the world—or rather, *worlds*—as existing in a state of perpetual flux. For Henri Bergson, mobility and change constitute the very *substance* of reality, which is why he felt that Classical substantialist ontology—according to which a sub-stance persists, immutable, *below* change—should be replaced by an ontology of duration. Rather than say, "a child becomes a man," we should say, "There is becoming from child to man." That way, it is the reality of becoming which is emphasised. "It is the only reality: childhood and manhood are no more than virtual halts, intellectual constructs" (Bergson, 312).

From such a perspective, there is no sense in defining bodies in terms of organic structure and anatomical form; in fact, there is no sense in defining bodies at all, for as much as definition implies imposing ends and edges (*fines*). Taking their inspiration from Bergson, Uexküll, and Spinoza, Deleuze and Guattari prefer to ask, what can a body do? In *A Thousand Plateaus*,

> a body is defined only by a longitude and a latitude: in other words the sum total of the material elements belonging to it (longitude); the

> sum total of the intensive affects it is capable of at a given power or degree of potential (latitude). Nothing but affects and local movements, differential speeds.
>
> (Deleuze and Guattari, 260)

Affects are becomings precisely because they refer to the passages and transitions between bodies:

> We know nothing about a body until we know what it can do, in other words what its affects are, how they can or cannot enter into composition with other affects, with the affects of another body, either to destroy that body or to be destroyed by it, either to exchange actions and passions with it or to join with it in composing a more powerful body.
>
> (Deleuze and Guattari, 257)

Ovid, in his *Metamorphoses*, already thought of bodies in this way (2001, 130). When he addresses the question of gender difference in the fable devoted to Tiresias, he doesn't define gender in terms of anatomy or even physiology. Besides, ovaries, the uterus, and mammary glands define women as mothers rather than as women per se. No, the question which allows Ovid to differentiate women from men is, famously: which of the two, male or female, feels greatest pleasure during sexual intercourse? Here, the criterion is the degree of intensity and a capacity: the capacity of being more or less *affected* (for "passivity" is as much a modality of power as is "activity"). The difference between two intensities of feeling is not the same as that between two weights or measures. You cannot totalise two degrees of intensity, just as you cannot add together two speeds or two temperatures. Along the scale of feeling, each difference is a difference in kind and not only one of degree. A change in affective capacity brings about a change in nature. This means that if a man does become capable of experiencing the kind of sexual intensity that Tiresias ascribes to women, he effectively and affectively becomes a woman, as was the case with Tiresias. Variations in our powers of feeling and acting are just as many becomings. Classical and modern fabulists, artists and film-makers make these becomings manifest by presenting actual bodily metamorphoses. However, the subjacent becomings are nonetheless real for having been fictionalised.

What, then, is the medium of affective metamorphosis? We have come to the third of Charles Foster's presuppositions: his failure or refusal to recognise the potentially beastly nature of human language, in general, and of literature, in particular. Foster's contention that Baker's becoming-peregrine is no more than wishful thinking is founded upon what is apparently incontestable evidence. If Baker really does become-peregrine, or at least if he starts behaving like one, why doesn't he steadily lose the power of human speech? On the contrary, his writing seems to be increasingly imbued with a special power and strangeness: "wingless nouns stoop and glide; burrow-dwelling verbs somersault on the edge of the atmosphere; adverbs behave disgracefully." Foster loves the strangeness, but he claims that it teaches him "more about language than about peregrines" (Foster, 6). But the very terms in which he couches his denial cannot but hint at the possibility of there being a continuity between the "symbolism of indivision" that Merleau-Ponty perceives in the forms, colours, and gestures of animals and the "exact symbolism" which characterises conventional human languages: the parts of speech behave like falcons; the rhythm of the sentences mirrors the movement of the bird.

Deleuze, for his part, doesn't dwell on the mimetic and rhythmic dimensions of language but insists upon the special nature of literary and artistic endeavour itself. He summarises his thought about the purpose of literature in a quote from D.H. Lawrence: "to leave, to leave, to escape . . ., to cross the horizon, enter into another life" (Deleuze and Parnet, 27). In *The Peregrine*, the bird opens up new styles of perception and affection for the narrator, a new "sensorium." As he pursues the falcon through its hunting fields, Baker enjoys an intensity of movement and affect unbeknown to his human self, but the revelation brought about by his enhanced perceptual capacities is entirely dependant, if we are to believe Deleuze, upon the act of writing:

> the problem of writing isn't separate from the problem of *seeing* and *hearing*: when another language is created in a language, language in its entirety tends towards an "a-syntactic," "a-grammatical" limit which communicates with its own outside. The limit is not outside of language, it is its outside: it is made up of non-verbal visions and auditions, but which only language can render possible.
>
> (Deleuze, 9)

This is why becoming-animal is not something that is simply recorded in literature or represented in art. It is effectuated, really and truly effectuated, by the extra-linguistic powers of language itself, that special kind of explicit symbolism that draws its power from the tacit symbolism of animal life.

A poem by the Englishwoman Steevie Smith (1902–1971), one of the most original and least easily classifiable writers of the twentieth century, exemplifies the preceding ideas to an uncanny degree. "Pretty" is a poem about a word and about the mode of existence of which this word is an expression.

> Why is the word pretty so underrated?
> In November the leaf is pretty when it falls
> The stream grows deep in the woods after rain
> And in the pretty pool the pike stalks

He stalks his prey, and this is pretty too,
The prey escapes with an underwater flash
But not for long, the great fish has him now
The pike is a fish who always has his prey

And this is pretty. The water rat is pretty
His paws are not webbed, he cannot shut his nostrils
As the otter can and the beaver, he is torn between
The land and water. Not "torn", he does not mind.

The owl hunts in the evening and it is pretty
The lake water below him rustles with ice
There is frost coming from the ground, in the air mist
All this is pretty, it could not be prettier.

Yes, it could always be prettier, the eye abashes
It is becoming an eye that cannot see enough,
Out of the wood the eye climbs. This is prettier
A field in the evening, tilting up.

The field tilts to the sky. Though it is late
The sky is lighter than the hill field
All this looks easy but really it is extraordinary
Well, it is extraordinary to be so pretty.

And it is careless, and that is always pretty
This field, this owl, this pike, this pool are careless,
As Nature is always careless and indifferent
Who sees, who steps, means nothing, and this is pretty.

So a person can come along like a thief—pretty!—
Stealing a look, pinching the sound and feel,
Lick the icicle broken from the bank
And still say nothing at all, only cry pretty.

Cry pretty, pretty, pretty and you'll be able
Very soon not even to cry pretty
And so be delivered entirely from humanity
This is prettiest of all, it is very pretty.

(Smith 1994, 274)

Strictly speaking, the word "pretty" belongs to the register of aesthetic judgement, but its very repetition and its increasingly exclamatory nature emphasize its inadequacy as a means of accounting for the intricacies of aesthetic experience. "Pretty" is a term like "nice," equally devoid of any real semantic content: we use them for want of a better word and, in using them, fall woefully short of the kind of elaborate verbal expression we would expect from a poet. Given that words can "become" just as living organisms can, what becomes of the word "pretty" in the poem "Pretty"? In being applied without distinction to all manner of natural phenomena that the poet perceives as she goes for a walk through the woods on a cold winter's evening, "pretty" should have the effect of levelling and deadening that experience by its constant reiteration. However, it has quite the opposite effect. This is because, to begin with, it is applied to phenomena that are far from being what we would normally describe as "pretty": the act of predation, pikes and owls, water rats, falling leaves, a simple field. More importantly, the steady intensification of the word "pretty," its progress from adjective to comparative ("prettier"), and from comparative to superlative ("prettiest"), accompanies an equivalent transformation in the poet-narrator herself. "The eye abashes"; the subject is cowed by an overwhelming sense of intense prettiness. The furtive nature of her sensory immersion—"Stealing a look, pinching the sound and feel,/ Lick the icicle broken from the bank"—is an indication of her gradual assimilation among the other denizens of the wood. The carelessness and indifference of Nature produces in her a state of ecstatic rapture: "pretty" is no longer a word belonging to a semiological system but a cry, a bark, the sonorous manifestation of a previously unknown affect. Most strangely of all, this affect evolves into a new capacity: the ability "Very soon not even to cry pretty," the loss of all remnant of human speech which heralds complete deliverance from humanity. In "Becoming-Intense, Becoming-Animal, Becoming-Imperceptible," the poetical voice at one and the same time reduces language to almost meaningless redundancy and intensifies its expressive capacities to incandescence. The conventional symbolism of human language drops back to the bedrock of natural symbolism: "This is prettiest of all, it is very pretty."

References

Baker, John Alec (2011). *The Peregrine*, The Hill of Summer & Diaries: The Complete Works of J.A. Baker. Edited by John Fanshawe. London: Collins.

Deleuze, Gilles (1993). *Critique et Clinique*. Paris: Minuit.

Deleuze, Gilles, and Guattari, Félix (1980). *Mille Plateaux*. Paris: Minuit.

Deleuze, Gilles, and Parnet, Claire (1977). *Dialogues*. Paris: Flammarion.

Descola, Philippe (2005). *Par-delà nature et culture*. Paris: Gallimard.

Foster, Charles (2016). *Being a Beast*. London: Profile Books.

Kohn, Eduardo (2013). *How Forests Think: Towards an Anthropology Beyond the Human*. Berkeley: University of California Press.

Lawrence, D.H. (1998). *Selected Critical Writings*. Oxford: Oxford University Press.

Merleau-Ponty, Maurice (1995). *La Nature. Notes de cours du Collège de France*. Edited by Dominique Séglard. Paris: Seuil.

Ovide (2001). *Les Métamorphoses*. Edited by Danièle Robert. Arles: Actes Sud.
Smith, Steevie (1994). *Collected Poems*. London: Faber & Faber.
Uexküll, Jakob von (2010). *Milieu animal et milieu humain*. Translated by Charles Martin-Freville. Paris: Payot & Rivages.
Viveiros de Castro, Eduardo (2009). *Métaphysiques cannibals*. Translated by O. Bonilla. Paris: PUF.

14 Cameras That Pose as Animals

Imagining Non-human Animals Through the POV Shot

Concepción Cortés Zulueta

1. Introduction[1]

A point-of-view shot (POV shot) is a cinematographic structure in which the camera shows what a character is looking at and may also move like that character. When the said character, the gazing-subject, is a specific non-human animal, the POV shot can be used to think about how certain images fluctuate between our symbolic projections and understandings on and about animals and their actual presence. In this chapter, my aim is to reflect on how POV shots try to convey the perceptual worlds and perspectives of non-human animals and on the evolution of the main uses of these filmic forms. This will approach the tensions and resistances between our human projections and meanings, and the perceptions and agencies of animals themselves—tensions that appear when the gazing sources are non-human animals.

2. The Origins and Development of the Classic Point-of-View Shot

In order to establish the basis of my analysis, I will first address the definition, origins, and early development of the POV shot as such. Through this approach, I will begin to point out and contextualize certain absences and limitations that affect the non-human animal POV shot.

Although there seem to be as many definitions as authors (Branigan 1984; Dagrada 2015; Gaudreault 1988; Mamer 2009, 10), the classic POV shot is usually outlined as a unit with two basic elements or shots. Thus, at times it is rightly called POV structure or pattern (Brewster 1982). One shot establishes the origin of the gaze—who is looking—and the other, the actual POV, takes the position of the gazing-subject and shows what is being looked at. A third additional and complementary shot can disclose the reaction of the character to what has been seen. This basic structure has experienced multiple variations, rearrangements, and deviations, as enumerated by Edward Branigan (1984, 111–19). There is some confusion or identification regarding whether the POV shot and the subjective camera (or shot), concepts which are considered more or less equivalent, are overlapping or discernible from each other. In fact, in Italian, French, and Spanish, the POV shot is usually called *soggettiva* (subjective), *caméra subjective* (subjective camera), or *plano subjetivo* (subjective shot) (Dagrada 2015, 15, note 2; Konigsberg 2004). Accordingly, there appears to be a certain tendency towards relating "POV" with optical or perceptual dimensions, and "subjective" with psychological or emotional ones.[2]

As a contemporary art historian, I am adopting a visual culture approach in order to address several films which include POV shots from the perspective of a non-human animal. My understanding of the POV shot will be intentionally loose at times. I will consider both POV and subjective shots—as I mentioned earlier, both are often translated in Spanish as *plano subjetivo*—as those that can be understood as emitted by a non-human animal character, even if some of them are not technically or theoretically orthodox.[3] Although Elena Dagrada points out the ambiguity of these terms and appears to slightly object to them (2015, 15, note 2), the fact that POV-subjective shots always allude to and mix both optical-perceptual and psychological-emotional dimensions, in my view, enriches this filmic form. My intention is to analyse these dimensions and to provide some overarching observations concerning the characteristics and functions of a cinematographic phenomenon that has not received enough attention,[4] despite being introduced by Jonathan Burt as "[o]ne of the most significant ways in which film explores the animal look" (Burt 2002, 54).

It is important to note that the POV shot is ingrained in the origins of film editing, as seen in the early film *Grandma's Reading Glass* (G.A. Smith, 1900). A child uses his grandmother's magnifying glass to enlarge several items (a newspaper page, pocket watch, caged canary, grandma's flickering eye, and a cat) which are shown framed by a circular black matte, as if it were the lens. Three of the items we see magnified are, in fact, animals (canary, cat, even Grandma) and have eyes themselves, with the potential to return the gaze. However, at this point non-human animals do not turn into gazing-subjects themselves, as if there were a certain resistance in this sense, although they appeared frequently in these early films.[5]

A widespread interpretation of the structure of *Grandma's Reading Glass* states that the POV shot here is a pretext which introduces those sudden close-ups, at a time when audiences could find cuts puzzling and difficult to understand. Dagrada, however, questions this assumption and considers the magnified bits as inserts or detail shots instead of close-ups (2015, 64), but she also stresses how the POV shot had a prominent role in early cinema's transition from an emphasis on optical tricks and illusions—what has been called a "cinema of attractions" (Gunning 1990)—towards linearization (Dagrada 2015, 141–87).

In fact, in his book on D.W. Griffith, Gunning analyses a POV structure from his third film (*The Red Man and the Child*, 1908) which is useful to illustrate this transition from "attractions" to narrative (1994, 70–4). In this POV pattern, the Native American protagonist learns that his gold has been stolen and his little friend abducted through the use of a telescope, so he proceeds to recover them. This sequence already displays the main functions of the classic POV shot, further developed in the following decades. On the one hand, attention is focused on something which allows the story to unfold by presenting or withholding information. We learn about the robbery and the abduction; we know that the leading character has *seen* and knows what has happened while those around him do not. On the other hand, the POV structure provides additional information about the gazing-subject, her or his subjectivity, perception, mood, reactions, and emotions. In Griffith's film this is accomplished through the character's reactions, but later on the POV shot could adopt some of the character's permanent or transitory perceptual peculiarities, or move like the character.

Among the many creative innovations of *Napoléon* (1927), Abel Gance's epic masterpiece, there is an early instance of this kind of embodied POV shot. The French filmmaker asked the cast to punch a camera in order to film a snowball fight, as if Napoléon, the child, was the one being hit. In Gance's words:

> [The camera] defends itself as if it were Bonaparte himself. It is in the fortress and fights back. It clambers on the wall of snow and jumps down, as if it were human. A punch in the lens. Arms at the side of the camera as if the camera itself had arms.
>
> (cited in Brownlow 1983, 57)

Therefore, in this snowball fight sequence, the camera takes the place of the character, a human animal. It merges with his movements and body, and this happens again in other parts of the film. So after *Napoléon*, the camera could show us what a gazing-subject was seeing and also embody that subject. In the following decades, therefore, Hollywood films from the classic era made extensive use of the POV shot, which appeared occasionally in many movies, especially in noir and horror films. However, other films experimented more thoroughly with it. *Lady in the Lake* (Montgomery, 1947) was filmed entirely from the point of view of its starring detective, whose face could only be seen in mirrors, and the first section of *Dark Passage* (Daves, 1947) was a long concatenation of POV shots, conveying the movements and emotions of Humphrey Bogart's character while he escaped prison: the strange angles due to his hiding places, the pace of his steps and nervous lookarounds, and so on.

Overall, POV shots began to codify novel situations. For example, the shot would fade to black if the character closed the eyes, become unsteady if drunk, fogged if smoking or about to faint, or jolted if running. A common embodied POV shot in horror and noir films was the stalking shot, when the camera insinuates that an unknown character is stalking another one from behind. Alternating with humans, spirits, or monsters, non-human animals also did the looking in some of these classic films and in some of their stalking shots, although, as I will detail, they tended to be portrayed or regarded as limited, rare, or surrogate gazing-subjects, revealing a certain resistance against imagining and considering them as full gazing-subjects.

For instance, going back to Gance's film, there is no embodied POV shot when the animal involved is not the human character, but Napoléon's pet eagle. The intertitle that introduces the bird as Bonaparte's companion is soon followed by a frontal comparison, and next, by a superposition of both their heads. Later, we see the child looking and, immediately afterwards, a shot representing his stare, showing the released and then recovered eagle perched in a tree and surrounded by a circular matte. The emphasis is on the bird's metaphorical connotations, on its association with imperial power. These human projections are reinforced in the final section of the film, where the eagle and its shadow lead the marching troops from above; again, no POV shot is emitted from this aerial position, no bird's-eye view is offered.

However, a contrast can be found in a much later film belonging to the genre of non-human horror movies, especially relevant due to the significance and prevalence of POV shots in Hitchcock's films. In *The Birds* (1963), a literal bird's-eye-view shot provided the aerial point of view of the feathered creatures. It is fixed above Bodega Bay while the gas station is on fire, more and more seagulls filling the screen and hovering over the town: an avian POV shot, although loosely understood, since the camera is not posing as a bird. Even if Hitchcock's original intention wasn't to provide the birds' perspective (Counts 1980), the view did destabilize the usual human-animal power relations, as if the seagulls were admiring from above the extent of the destruction they were causing.

3. Two Reflections on Non-human Animal POV Shots

Once the classic POV shot was established, the non-human animal POV shot began to occasionally appear. Yet it seems that animals were not so widely considered as potential gazing-subjects—as POV shot emitters—and when they were considered so, the resulting shots often carried connotations of menace and antagonism (Gregersdotter, Höglund, and Hållén 2015).

Subsequently, when animals other than humans act as the source of a POV shot, two reflections stand out, and I will build upon them in the following sections. First, addressing non-human animals as gazing-subjects and as agents able to act according to what they see is not always a given, as they were long considered as mechanical and/or instinctual beings, denied emotions and consciousness (Allen and Trestman 2016). In this sense, the film theorist Edward Branigan states in his influential book *Point of View in the Cinema* that a POV shot "requires the presence of a character, but more than a 'body' is necessary—the character must be 'aware' and 'looking' in a normal way" (1984, 108). However, immediately afterwards, he reduces the requirement to "a mere presence of awareness," since he finds that character is "a rather complex notion" (1984, 108) that he appears to be denying to other animals, as "much less [than a character] is needed to found a POV shot; for instance, the presence of an animal can give rise to a POV shot" (1984, chapter 5, note 13).

In a way, therefore, Branigan acknowledges the awareness of other animals but sees them as limited in other aspects. In my view, this stance simultaneously recognizes the cinematographic gaze of non-human animals, while also reflecting a certain resistance against considering their gaze a full one. Therefore, it aligns with what can be seen as lacking the conceptual imagination which might explain a certain scarcity or narrowing of non-human animal POV shots, as well as some of their peculiarities.

Likewise, this resistance can be related to the comings, goings, blends, and tensions between the two most common uses of the non-human animal POV shot I have detected so far. The first of these portrays the non-human animal as a menacing other, who stalks and attacks us. Instances of this menacing non-human animal are plentiful and can be interpreted in a myriad of ways: from the fear of the unknown and foreign to the anxiety of becoming prey while we still think we are the predators, to the uneasiness our inner animal may cause us (especially in the case of hybrids), and so on. On the whole, this portrayal as a menacing other limits the potentialities of non-human animals characters, reducing them to stereotyped dualities and to secondary and antagonistic roles (Gregersdotter, Höglund, and Hållén 2015).

The second use of the non-human animal POV shot would be to address non-human animals as subjects with their different and complex side of things—to treat them as full characters and agents. Thus, it is possible to address their perceptions, worlds, and modes of communication but also to anthropomorphize and distort them. Nonetheless, it is not uncommon to find both uses blended together to some extent, in different balances and emphases.

On the other hand, and as a second reflection, the information that a non-human animal POV shot can offer about the animal involved has many distinct layers compared to a regular, human POV shot. Despite the fact that a camera does not see as a human being will, we tend to assume that regular footage is like regular human vision. In Branigan's words: "we may say that the non-distorted, or in-focus, POV shot represents *no* special perceptual circumstances; that is, normal awareness" (1984, 108). Thus, what departs from the regular audiovisual circumstances can be read as temporary or permanent perceptual peculiarities of the gazing-subject, especially if it is a non-human animal—think about bird's-eye views and fisheye lenses. Therefore, a POV shot can be used to convey how the perception and perspective of the said animal differs from ours, either to misconstrue that perspective or to bring that animal world closer to us.

4. Non-human Animals as Limited, Surrogate, or Rare Gazing-Subjects

In line with how Branigan deals with animals as sources of POV shots, there tends to be a certain tension towards regarding them as full gazing characters, full cinematographic subjects. I will present some early and classic examples that manifest this tension in different ways.

One early precedent, though somewhat loose, can be found in *Clash of the Wolves* (Smith, 1925). In the first few minutes of this Rin Tin Tin film, the gaze of the wolf-dog, Lobo, scans the desert beneath him in search of prey, from left to right, while standing on the top of a rock. Lobo is both wolf and dog, an enemy that turns into a friend, and his mixed identity along with his transitions between wilderness and civilization are at the core of the plot. However, it is his aforementioned predatory gaze, which can be related to stalking shots, that grants him what most closely falls within the definition of a POV shot.

From the classic era onwards, many horror and monster films employed the POV shot in order to unsettle the viewers or to get them immersed in the action. *King Kong* (Cooper and Schoedsack, 1933) is a significant example of the use of non-human animal POV shots, as the giant ape can emit filmic gazes. This seems to be emphasized in his first encounter with Ann Darrow, when the camera zooms in to a close-up of King Kong's eyes, and then she is shown screaming,[6] and similarly later, when the gorilla is holding and inspecting her in his hand while he climbs a building, in a more canonical POV shot. Interestingly, Fay Wray's character is not able to do the same. During the screen test on the deck, Darrow looks directly into a camera that stands for her fictitious director, yet the corresponding shots do not "originate in her eyes," since we see the director slightly from the side. She is just the object of other gazes, coming either from King Kong or from a human male, never the source herself. Here, the distance between the gaze of King Kong and that of a male human is not so great; after all, both are primates, apes.

As I mentioned before, some early non-human animal POV shots were stalking shots, emitted by a pursuing character that could be an animal, an animal monster, or a hybrid. Moreover, although there are no POV stalking shots in the original *King Kong* film, they can be found in another gorilla film, *Bride of the Gorilla* (Siodmak, 1951), where they are very prominent.[7] The film begins and ends with a tracking shot from inside the jungle, which frames the stalking shots of the movie and insinuates that something is lurking within the vegetation. In the final section we can witness how the leading character, a human male metamorphosed into a gorilla, pushes aside branches and lianas using his hairy hands, while he stalks and advances towards his human girlfriend, all from his gorilla POV.

Though menacing or monstrous POV shots are prevalent among the animal ones, other early shots diverge from this association. Still, it is as if the animals involved were being addressed as surrogates, or exceptions. The film *Eyes in the Night* (Zinnemann, 1942) is a singular case. It tells the story of how the blind detective Duncan Maclain and his seeing-eye dog Friday defeat a squad of conspirators trying to steal a scientific formula. As underscored by Martin Norden, who analyzed a couple of the film's POV structures shot by shot, Zinnemann decided to build the blind man's point of view "by photographing much of the film from the seeing-eye dog's perspective" (1999, 62). That is, following "the classic style of Hollywood editing, but with a twist: a canine, not a human, does the looking" (1999, 63). Norden also quotes Branigan's observations on animals as sources of POV shots but stresses that "*Eyes in the Night*'s use of nonhuman point of view is considerably more involved than the simple form suggested by Branigan" (1999, 63).

Norden details a complex sequence—where the blind detective is hidden in a closet, and the dog under the bed, while watching an unaware conspirator—to point out a replacement of the looks, as one of the shots appears to be emitted by the dog, but from the blind detective's position. In Norden's view, "the canine is a veritable extension of Maclain himself," and "his gaze becomes Maclain's gaze" as "his perspective is at times a surrogate for his master's" (1999, 64). Thus the human character's visual prerogatives are transferred to his surrogate, the dog, who is able to emit gazes as if by contagion on behalf of his master. However, it is relevant to note that there are other canine sequences and POV shots in the film where humans are absent, like in the final exchange of glances between the seeing-eye German shepherd and his poodle crush, a humorous side story which is employed as a romantic happy ending. Still, it can be argued that this last sequence is secondary to the main plot, and even slightly "cartoonish."

A similar exchange of glances, but feline instead of canine, opens the experimental *The Private Life of a Cat* (Hammid and Deren, 1944)—made shortly after *Eyes in the Night*—a 22-minute circular film portraying the conception, birth, and upbringing of a litter of five kittens by their feline parents, the title of which seems an allusion to Julian Huxley's short documentary *The Private Life of the Gannets*[8](1934). Since the latter focuses on a colony of birds established on the rocky island of Grassholm, the allusion puts forward a suggestive contrast between the life cycles of wild and domesticated animals.

Hammid and Deren's film is entirely narrated from the cats' point of view, including the widespread use of POV structures from the female—a series of low-angle shots of cupboards and cabinets which match her moving gaze in search of a safe space to deliver her kittens—the male—another set of filmic gazes from him which follow the female while she moves the kittens in her mouth, one at a time—and a kitten—low angles of his feline father from the bottom of the scratching pole which he has just climbed. In Rosalind Galt's words, "[t]he editing is utterly classical, except the looks through which we look are cats' looks. . . . The camera takes up the cat's optical POV, an ordinary strategy in classical cinema but here made strange by asking the spectator to look *as a cat*" (2015, 49). Galt also stresses how the look disintegrates "in favour of an emphasis on texture, furriness and sensation" (2015, 49), thus partially questioning the human—and by extension, also cinematographic—hierarchy of the senses.

In my view, as it is framed as an experimental attempt, what prevails and is underlined is this experimental quality, despite the features of the film itself. It is as if at that point, those cats where exceptional, a rare instance of cinematographic gazing-subjects, but not an everyday occurrence. As if the message delivered could and tended to be read like "now picture cats like full gazing-subjects, imagine they can be so," although this was something only happening in experimental films. Thus, the roles of animals as menacing monsters, surrogates, or experimental characters again limit and define their position as POV shot emitters.

5. How Does a Fly See the World? Fly Perceptual Perspectives Through Entomological POV Shots

One of the two reflections about the non-human animal POV shot that I outlined earlier concerned how this filmic form could be used to convey the perception of an animal, either to misconstrue it or to bring us closer to that animal's world. I will approach this issue by turning to flies and to entomological POV shots, starting with a film that puts us in the position of a monstrous hybrid creature, partially human, partially non-human.

In *The Fly* (Neumann, 1958), a central sequence tries to represent how a giant hybrid-fly would see. When the scientist's wife removes the cloth covering her husband and exposes the head of a giant fly, the first regular close-up shows her screaming. The camera then zooms in to the fly's eyes, and the subsequent POV shot multiplies her image, accompanied by her second scream. This pattern is repeated with her third scream as she steps back. Thus, according to the POV shot, this fly sees a repeated image all over its field of view, as if each facet of its compound eyes perceived the scene separately and the brain composed a mosaic with them.

Of course, this cinematographic fly is a human-insect monster created by radiation, and this mosaic multiplication recalls other sequences present in surrealist and avant-garde films, inherited by dream or hallucinatory sequences. This surreal touch reinforces the disturbing effect of the images: the radioactive monster manages to raise common fears towards nuclear power elicited by the Cold War. Moreover, this visual allusion to how a fly—not just a giant hybrid fly but any fly—would see us made the 1950s audience feel like strangers in their own homes, watched by dozens of compound insect eyes, their faces multiplied by each one of them. As a side note, this mosaic-like POV shot is not reserved just for flies: in *Phase IV* (Bass, 1974), the ultra-intelligent ants who challenge humanity briefly gaze at humans in this compound manner.

Despite its horror movie context, a certain informative residue remains in *The Fly*, derived from earlier films that attempted to convey how a house fly sees. In fact, the 1915 newspaper review of one of Charles Urban's educational and scientific films reads:

> Up-to-dateness is further exemplified by a number of pictures entitled "The menace of the fly" . . . by the far [sic] the most extraordinary part of this film is a section of photographs taken through the lenses of a fly's eye, thus giving an idea of how the fly sees his mortal enemy—man.
>
> (*The Daily Telegraph*, May 27 1915, as quoted in Burt 2002, 124)

According to the reviews of *The Menace of the Fly*,[9] the British Pathé short called *Eyes—By The Thousand!* (1930) closely resembles the fly-optics approach of Urban's film and might be even copying it.[10] The latter praises the fly's eyes as "the sharpest and best in all creation" and proceeds to show enlargements of their multiple lenses. Then, "[a]n experiment in optics" follows, comparing regular camera vision with fly vision: how we see a moving automaton clown, as photographed through "an ordinary cine-lens"—probably standing for human vision—followed by how we see it through 50 of the fly's lenses, obviously filmed with another cinematographic lens themselves. The automaton clown is multiplied and its tiny replicas fill the screen, one for each fly lens. The layout of the image is like *The Fly*'s POV shot, and both Urban's and Pathé's educational films share, with the 1958 horror movie, a comparable emphasis on the menace posed by flies. Although, in the former, this menace is related to public health concerns—linked to the role of flies as disease carriers, and their exponential multiplication[11]—while the main goal of the horror film was to unsettle the viewers.

There is also an intriguing parallel between these fly films and the images published by the biologist Jakob von Uexküll (1864–1944) around the first half of the 1930s to illustrate his notion of *Umwelt*, the surrounding sensory world of every living creature. Uexküll's images show a street as seen by a human and by a fly, given their disparate visual acuity, that is, with much less visual information in the fly's case, but without any mosaic or image multiplication. He explains the method he had developed to obtain the coarser images, interposing a lattice over a regular photograph, which was taken as an equivalent to human visual acuity (Von Uexküll and Kriszat 1957, 20–1). Thus, Uexküll also displayed a regular photograph as comparable to human vision and modified it in order to illustrate certain characteristics of different perceptual worlds. There is likewise a shared educational impulse, and practice, between Uexküll's images and the fly films, even if the results of the latter were not as factual as intended. Uexküll was actually only illustrating a very specific aspect of vision across species, while the fly films made an incorrect assumption about how the eyes of a fly worked, based on their external optics and compound appearance.

Regarding the influence of these representations, their strength and capacity in determining our common knowledge of insects is worth considering; how easily they are disseminated, assumed, and remembered, even embodied, once they have been read and understood as illustrations of how other creatures see. In fact, mosaic sequences like the one in *The Fly* may explain why so many people still think that insects see like this or why some supposedly scientific toys propagate this kind of bug vision. Some filmmakers have tried to counter this belief: in another POV sequence in the film *Microcosmos* (Nuridsany and Pérennou, 1996), approaching a bee's perception of a field, no multiplication is shown at all, only bits of reduced visual information which compose a whole image. Interestingly, this sequence has tended to be poorly understood and was received with incomprehension and disapproval,[12] as noted by Evans (115–16).

Nonetheless, in spite of the fact that we can and will misconstrue other animals and their perspectives, projecting upon them our misconceptions even when we are trying to be scientific, I believe the use of pictures to convey how other animals perceive can prove useful, in different ways and degrees, in order to imagine non-human animal worlds and perceptions, as I have argued elsewhere (Cortés Zulueta 2013).[13] Similarly, Uexküll employed these images himself and also indicated the potential of painters to codify the worlds, or *Umwelten*, of other animals into sensory terms more understandable in human ones:

> [N]othing can be more instructive for a biologist who wants to analyse the [perceptual] notes of things than turning to painting and treating painters who, as nobody else, have to delve into the perceptual world in order to reproduce it in a picture. . . . Nowadays, to manage with three perceptions is considered as the highest art; something that can also be done by sea urchins. Of course, for biologists these simplifications of the panorama of [perceptual] notes are of the utmost

interest, because they allow the immersion of the gaze deeply inside the perceptual world of animals.

(1922, 73–4)[14]

6. A Shift in the Tensions Between Non-human Animals as Menacing Others and as Full Gazing-Subjects

In a previous section, I explained how POV shots portrayed non-human animals as menacing and limited gazing-subjects in early and classic cinema. In later decades, the tensions between those two main conventions evolved and shifted. One way to show this—probably among others[15]—is to turn towards a reactivation of animal stalking shots, considerably influenced by a famous shot featured in Spielberg's *Jaws* (1975). Almost everyone can picture the underwater sequence with its tuba leitmotif, showing the swimmers from the POV of a shark choosing its prey. This shot reveals a liquid world, embodies the eye—just a single eye—that may be peering at us from the depths, and the uneasiness which comes with that acknowledgment. After *Jaws*' success, other similar films kept stalking-animal and POV shots as their key elements, either on the oceanic side—as in *Orca* (Anderson, 1977)—or on the canine side—like in *El perro* (Isasi, 1977) or *White Dog* (Fuller, 1982). Both of the latter are political commentaries, on prosecution in Latin American regimes and racism towards African-Americans in the United States, respectively: again, human projections onto those portrayed as real animals. Fuller's film is particularly interesting, because it was a mixture of elements from monster movies like *Jaws*—the initial idea, even before Fuller was appointed as its director, was to make a "*Jaws* on paws" (Dombrowski 2008, 191)—since the dog embodies the racism of those who trained him to attack black men, while he remains an actual subject and character, and a friendly companion for the white female protagonist. In Fuller's film, the dog is both a stereotyped menacing and an attacking other, as well as a full gazing-subject. However, the POV shots tip the scale towards the latter, since they tend to describe the dog's perspective, position, and thinking processes—his conversation-like interactions with human characters, his reactions to reconditioning sessions, and his attempts to escape from his enclosure. In this sense, Fuller later stated in an interview how, through the editing, he had tried to build the dog's character as a thinking and gazing-subject:

> I directed him like a human being. . . . I talked to him like an actor. Did you notice the close-ups of his eyes in the film? I looked him right in the eye. I wanted the dog to be shown as a thinking being. Me, I hate movies like *Lassie*.
>
> (Peary 2012, 90)

A similar combination of friendliness and menace appears again in *The Simpsons*' episode *Bart's Dog Gets an F* (1991), where the family dog has to be educated, and which includes several scenes from his point of view. Though animations, they act as a comment on the two most common uses of the non-human animal POV shot: the previously mentioned menacing other and full gazing-subject characterizations. However, in *The Simpsons*, the menacing POV shot is treated humorously, as we see what the family dog is seeing while he attacks the kitchen and eats Homer's giant cookie. Therefore, here the emphasis has shifted from depicting the non-human animal as a menacing other to showing it as a friendly and naughty companion, a family member which is also a full character and gazing-subject, as it partially happened in *White Dog*.

Accordingly, in *The Simpsons*' episode, animated sequences in the form of POV patterns are intended to express Santa's Little Helper's canine side of things, even perceptually. The images are almost black and white but have some blue-and-green hues, which reveal an imperfect attempt to convey that dogs are bichromatic and able to see colours, though less than the average trichromatic human. The climax of the story arrives when the dog begins to comprehend some words, as the gibberish we were hearing before, during the POV shots, is suddenly replaced by some human speech, something that reveals that the dog is now able to understand and follow Bart's commands, in accordance with canine abilities. On the whole, this animated episode is a further step forward concerning the tensions and negotiations between the two main uses of the non-human animal POV shot. It reverses what was common in early and classic cinema, humorously undermining the "menacing other" of animal characterizations while reinforcing them as "full gazing-subjects."

It may seem paradoxical that, as seen in *The Simpsons*' episode, a cartoonish dog that is often anthropomorphized—like when he is depicted smiling—manages to convey the canine side of things better than some live action films, in spite of the presence of actual animals in the latter. This is a further instance of the comings and goings between non-human animals themselves and their portrayals. Non-human animal POV shots are always representations, human accounts of other animals and their worlds. In them, humans can grant or deny those animal characters a gaze, or agency, either bringing other animals closer to our understanding and imagination or distorting them.

It is important, however, to keep in mind that this agency is a bestowed one, constructed using cinematographic conventions, and it is not the agency of the filmed and actual animals themselves. This distinction may clarify why sometimes, an animated animal like *The Simpsons*' dog, can be considered a better depiction of a full gazing-subject than a filmed, live one. On the other hand, it is worth noting that the presence of animals in films can also contribute towards making us more aware of their perspectives, since we cannot obscure that presence entirely.[16]

7. Animal Resistances

Since filmed animals retain their bodily presence and at least some of their own agency and mood, they can still convey things unrelated to what human directors and viewers require from them. At times, you can even see a happy dog in a tense and violent scene, as in Herzog's *Bad Lieutenant: Port of Call New Orleans* (2009), where the dog seems to be playful with the men threatening him. After all, the humans involved are pretending, and the dog probably knows the difference.[17]

Along this line, a similar contrast or friction can also appear in non-human animal POV shots. At times, these filmic patterns reveal certain animal resistances, non-human animals opposing our meanings and projections. If we analyse films where horses dialogue with humans through shot/counter-shot patterns—which can be considered an extension or variation of POV structures—the conventional cinematographic codes are compromised. Horses do not talk as such, but this can be circumvented, as not all filmic dialogues are spoken ones, and other bodily expressions can convey the horse's feelings or reactions, like gazes, or movements of the tail, legs, head, ears, and eyes. The main problem is that horse heads are different from human heads—horses have their eyes on both sides of their head—making it difficult to record a frontal shot of a horse and to capture a gaze this way. In fact, searching "horse" and "face" in Google Images will collect many horse profiles, not only frontal views. Thus, cartoon horses tend to be anthropomorphized, and their eyes migrate to the front of their heads in order to evade the problem, while in non-animated films, the human character often finds himself or herself talking with a profile, or even with a close-up of a single horse eye, instead of with a full face or gaze.

In fact, Burt has noted how POV shots fetishize the eye, "and often it is the single eye rather than the pair" (2002, 54). He has also underscored how this tendency punctuates *The Horse Whisperer* (Redford, 1998), a film with several POV shots from Pilgrim, a troubled equine. What Burt does not mention is the potential role of the horse anatomy in this case, or in the alternation between images of two-eyed (human) and single-eyed (horse) gazes in the human-horse filmic dialogues featured in the movie. Moreover, horses are not alone in this sense, as this emphasis on the single eye extends to other animals like sharks (*Jaws*) or cetaceans (*Orca*) due to their respective anatomies. This acts as a reminder of how cinematographic forms are not natural and universal but rather anchored in our human world and cultures, in our perception and bodies. Thus, when we try to project and impose certain meanings onto other animals through cinema—animals with bodies other than human—some frictions and oppositions emerge. The presence of other species, therefore, challenges cinematographic foundations and conventions, as shown by filmic equine dialogues.

8. Coda: Non-human Animals as Careless Witnesses to Human Dramas

There is a more recent use of the non-human animal POV shot, in which non-human animals act as transitory witnesses of human actions. I see this kind of shot as a way to allude to how other animals can be temporarily intrigued by human dramas, but otherwise indifferent to them; it is as if they take brief notice, but ultimately could not care less. This recent employment can be framed as a response of certain filmmakers to all those human meanings projected onto animals, maybe even as an acknowledgement of their resistances and as an attempt to balance the weight granted to human stories by noting the existence of non-human ones.

In Almodóvar's *Qué he hecho yo para merecer esto* (1984), we are shown how a man hits a woman from the point of view of a lizard, who witnesses the blow and gets stained with some drops of her blood before disappearing. In Gondry's *Human Nature* (2001), a raven wonders and watches how the climax of the film unfolds, providing a circular narrative to the story. In Herzog's *Bad Lieutenant: Port of Call New Orleans* (2009), the camera poses as a crouching or creeping creature standing on the side of the road after a car accident involving an alligator before breaking the illusion in the same shot, revealing the position of another alligator looking at the scene[18]: the next shot follows the second alligator as the reptile leaves. Thus, Herzog manages to use and challenge the POV structure simultaneously. All these films refer to these careless witnesses, recalling how many animals behave around us. It is refreshing as it briefly removes the spotlight from humans.

All these movies tell human stories from a human perspective, but in what I consider a humbling turn—though mostly ephemeral—they briefly track the other gazing eyes and points of view around us, alluding to millions of other stories where we are neither protagonists nor relevant. In the end, this recent use can be framed as a consequence of the emphasis on animals other than human as fully gazing and perceiving subjects. After all, if we recognize that animals have a perspective of their own, the next step is to acknowledge that this perspective does not necessarily have to revolve around humans, and this impacts the anthropocentric foundations of cinematographic codes and conventions. Somehow, I believe that we can appreciate all these other stories and worlds beyond every non-human animal POV shot if we use it to reflect on the comings and goings of this filmic structure between human projections and meanings, and animals themselves.

Notes

1. This chapter was finished while I was a Juan de la Cierva postdoctoral fellow of the Spanish Ministry of Science, Innovation, and Universities at the Art History Department, University of Málaga, and a member of the Spanish National I+D project Arte y Cognición Corporal en los Procesos de Creación: Sensibilización Ecológica del Yo en el Entorno (HAR2017–85485-P).
2. See how Galloway reads Branigan (Galloway 2006, 42–6), or Dagrada's remarks (2015, 15, note 2).
3. With this, I mean that some of the POV shots I address might not be strictly emitted from the position or angle of the character they supposedly belong to or may have been identified as subjective shots instead of POV shots by some authors, etc.
4. Branigan acknowledges that a POV shot can be also non-human or inhuman, like that of animals, dead men, invisible spirits, computers, statues, and aliens (1984, 104–5) and labels those POV shots as "deviant" (1984, 221). Galloway briefly refers to certain predatory animals among the gazing-subjects whom he identifies as protagonists of the majority of subjective shots in cinema, among them aliens, criminals, monsters, and other inhuman characters.
5. Some of them were early nature and animal films, like the ones produced by Charles Urban (1867–1942). Among these was *The Cheese Mites* (1903), a microscopic POV film that made those tiny creatures visible. It was accompanied by several more, devoted to flies and other insects, one of which I will address in the following.
6. Here, Fay Wray is not shown from above, so this shot does not match the position of King Kong's eyes.
7. Another interesting example of an animal stalking shot is *Cult of the Cobra* (Lyon, 1955), mentioned by Branigan as an instance of an animal (stalking) POV shot, where the image is elliptically distorted to convey the gaze of the cobra.
8. About this film, see Mitman (2013, 76–9).
9. For another review, which describes how *The Menace of the Fly* shows the stages of the fly's growth, see *The Daily Mail*, November 10, 1916.
10. I wonder if it could be a fraction of Urban's film, since the British Pathé collection has other "micro" films and footage, either produced by Urban or directed by Percy Smith. The film is available for preview on the British Pathé website: www.britishpathe.com/video/eyes-by-the-thousand/query/thousand+eyes
11. Such concerns were behind didactic books and control efforts dating from the beginning of the century, like the ones by the entomologist Leland Ossian Howard. See Howard (1911).
12. I am grateful to Éric Baratay for sharing with me the producer Jacques Perrin's dissatisfaction with how poorly understood *Microcosmos*'s apian POV shot was, prompting me to look into this issue.
13. In fact, in the referenced chapter, I discussed the limitations, flatness, perceptual assumptions, and sensory hierarchies of this kind of picture through the practices of several contemporary artists (Cortés Zulueta 2013).
14. I translated Uexküll's quote from a Spanish translation. I am not aware of an English translation.
15. This approach allows us to contrast and compare the shift in the tensions between both uses and characterizations in the way they are present simultaneously, in Fuller's *White Dog* (1982) and in *The Simpsons*' episode (1991), which I analyse. Instead of using the reactivation of stalking shots and menacing animals as a starting point, however, it would be also interesting to focus

on other films with POV shots that addressed animals as gazing-subjects during the sixties and seventies, like Bresson's *Au Hasard Balthazar* (1966), Disney's *That Darn Cat!* (Stevenson 1965), or *Benji* (Camp, 1974), a film mentioned by Branigan as an example of an animal POV shot.

16. For an interesting reflection about what it means for an animal to "act" in the filmic confluence between the mutual gaze of human and animals, see Burt (2002, 32).
17. I wouldn't be surprised if Herzog welcomed this contrast between the acting humans and the dog just being himself.
18. I owe to Valeria Camporesi the inclusion of *Bad Lieutenant*'s alligator POV shot.

References

Allen, Colin, and Trestman, Michael (2016). "Animal Consciousness." In *The Stanford Encyclopedia of Philosophy*, edited by Edward N. Zalta. Metaphysics Research Lab, Stanford University. https://plato.stanford.edu/archives/win2016/entries/consciousness-animal/.

Branigan, Edward (1984). *Point of View in the Cinema: A Theory of Narration and Subjectivity in Classical Film*. Berlin/New York: Mouton.

Brewster, Ben (1982). "A Scene at the 'Movies'." *Screen*, no. 2: 4–15.

Brownlow, Kevin (1983). *Napoleon, Abel Gance's Classic Film*. London: J. Cape.

Burt, Jonathan (2002). *Animals in Film*. London: Reaktion Books.

Cortés Zulueta, Concepción (2013). "How Does a Snail See the World? Imagining Non-Human Animals Visual Umwelten." In Carla Taban, *Meta- and Inter-Images in Contemporary Visual Art and Culture*, 263–79. Leuven: Leuven University Press.

Counts, Kyle B. (1980). "The Making of Alfred Hitchcock's the Birds: The Complete Story Behind the Precursor of Modern Horror Films." *Cinefantastique*, no. 12 (Fall): 15–35.

Dagrada, Elena (2015). *Between the Eye and the World*. Bruxelles: Peter Lang.

Dombrowski, Lisa (2008). *The Films of Samuel Fuller: If You Die, I'll Kill You!* Middletown, CT: Wesleyan University Press.

Evans, Georgina (2015). "A Cut or a Dissolve? Insects and Identification in Microcosmos." In Michael Lawrence and Laura McMahon, *Animal Life and the Moving Image*, 108–20. London: Palgrave Macmillan.

Galloway, Alexander R. (2006). "Origins of the First-Person Shooter." In *Gaming: Essays on Algorithmic Culture*, 39–69. Minneapolis: University of Minnesota Press.

Galt, Rosalind (2015). "Cats and the Moving Image: Feline Cinematicity from Lumière to Maru." In Michael Lawrence and Laura McMahon, *Animal Life and the Moving Image*, 42–57. Paris: Méridiens Klincksieck.

Gaudreault, André (ed.) (1988). *Ce que je vois de mon ciné. La représentation du regard dans le cinéma des premiers temps*. Paris: Méridiens Klincksieck.

Gregersdotter, Katarina, Höglund, Johan Anders, and Hållén, Nicklas (2015). "Anthropomorphism and the Representation of Animals as Adversaries." In *Animal Horror Cinema*, 206–23. Basingstoke: Palgrave Macmillan.

Gunning, Tom (1990). "The Cinema of Attractions: Early Film, Its Spectator and the Avant-Garde." In Thomas Elsaesser, *Early Cinema: Space, Frame, Narrative*, 55–62. London: British Film Institute.

Gunning, Tom (1994). *D. W. Griffith and the Origins of American Narrative Film: The Early Years at Biograph*. Urbana/Chicago: University of Illinois Press.
Howard, L.O. (1911). *The House Fly, Disease Carrier: An Account of Its Dangerous Activities and of the Means of Destroying It*. New York: Frederick A. Stokes Company.
Konigsberg, Ira (2004). *Diccionario técnico Akal de cine*. Barcelona**: Akal.
Mamer, Bruce (2009). *Film Production Technique: Creating the Accomplished Image*. Belmont, CA: Wadsworth Cengage Learning.
Mitman, Gregg (2013). *Reel Nature America's Romance with Wildlife on Film*. Seattle: University of Washington Press.
Norden, Martin F. (1999). "The Eyes Have It: Dimensions of Blindness in Eyes in the Night (1942)." In Arthur Nolletti, *The Films of Fred Zinnemann: Critical Perspectives*, 55–68. Albany: State University of New York Press.
Peary, Gerald (ed.) (2012). *Samuel Fuller: Interviews*. Jackson: University Press of Mississippi.
Von Uexküll, Jakob (1922). *Ideas Para Una Concepción Biológica Del Mundo*. Madrid: Calpe.
Von Uexküll, Jakob, and Kriszat, Georg (ill.) (1957). "A Stroll Through the Worlds of Animals and Men: A Picture Book of Invisible Worlds." In Claire H. Schiller, *Instinctive Behavior: The Development of a Modern Concept*, 5–80. New York: International University Press.

Films

Au Hasard Balthazar (Bresson, 1966).
Bad Lieutenant: Port of Call New Orleans (Herzog, 2009).
Bart's Dog Gets an F (Groening, 1991).
The Birds (Hithcock, 1963).
Bride of the Gorilla (Siodmak, 1951).
The Cheese Mites (Urban, 1903).
Clash of the Wolves (Smith, 1925).
Cult of the Cobra (Lyon, 1955).
Dark Passage (Daves, 1947).
El perro (Isasi, 1977).
Eyes—By The Thousand! (British Pathé, 1930).
Eyes in the Night (Zinnemann, 1942).
The Fly (Neumann, 1958).
Grandma's Reading Glass (G. A. Smith, 1900).
The Horse Whisperer (Redford, 1998).
Human Nature (Gondry, 2001).
Jaws (Spielberg, 1975).
King Kong (Cooper and Schoedsack, 1933).
Lady in the Lake (Montgomery, 1947).
The Menace of the Fly (Urban, 1915?).
Microcosmos (Nuridsany and Pérennou, 1996).
Napoléon (Gance, 1927).
Orca (Anderson, 1977).
Phase IV (Bass, 1974).

The Private Life of a Cat (Hammid and Deren, 1944).
The Private Life of the Gannets (Huxley, 1934).
Qué he hecho yo para merecer esto (Almodóvar, 1984).
The Red Man and the Child (Griffith, 1908).
That Darn Cat! (Stevenson, 1965).
White Dog (Fuller, 1982).

15 Playing (With) the Non-human

The Animal Avatar in *Bear Simulator*

Michael Fuchs

In the first paragraph of his book *Parables of the Posthuman: Digital Realities, Gaming, and the Player Experience*, Jonathan Boulter wonders, "In what ways does gaming alter our perception of the human subject?" (2015, 1). This opening question implies that video games make a central promise to players—the experience of becoming more-than-human. Recent video games in which players control non-human animals have drawn on this idea in their promotional paratexts. For example, the official website of the free video game *I, Predator!* (Kuma Games 2011), which was co-produced by Animal Planet, promises players that they will "[b]ecome one of nature's real life [*sic*] predators" ("I, Predator" 2009) while playing the game. This tagline, in fact, suggests that video games such as *I, Predator!* may open up spaces of animal becoming.

In his discussion of the role of non-human animals in video games, Adam Lindberg takes an opposite stance, maintaining that digital games "formalize problems of mastery" of both the animalistic traits inside all humans and, more generally, the non-human Other—video games "are the formal apotheosis of humanism's central conflict: the separation of the human from the animal" (2017). From this perspective, video games are prime examples of the "*unprecedented* proportions of . . . the subjection of the animal" characteristic of contemporary Western society (Derrida 2008, 25; italics in original).

However, in one way or another, all games—no matter whether digital or analogue—make such mastery impossible. After all, "[t]he attraction of a game, the fascination it exerts, consists precisely in the fact that the game masters the players," Hans-Georg Gadamer concluded decades ago (2013, 111). As much as players may try to master the game, they simply cannot, as any game is composed of underlying algorithms players do not have access to. The non-human game accordingly controls the human—human actions, in particular.

This human urge for mastery, for dominion of the planet, is simultaneously a remnant of the past and an idea deeply entrenched in human thinking (especially Western thinking, of course). Likewise, Western civilization still clings to the idea of what Fredric Jameson so aptly described as "the bourgeois ego, or monad" (1991, 15)—that is, the notion of a coherent and stable human subject and identity. Similar to how video-gaming challenges the idea of human mastery of the non-human, it also highlights the incoherence and instability of both living and breathing human beings and the category of "the human." Tellingly, in *Avatar Emergency*, Gregory Ulmer highlights that when users create avatars for the virtual spaces they roam, they "put [their] sel[ves] into the prosthesis of the Internet" (2012, xi), as the users' beings fuse with their online avatars, creating a new being—a new identity. Likewise, during gameplay, players do not simply control and master their onscreen personas but interact with them in a variety of ways. Accordingly, in this chapter, I will argue that animal avatars provide potential vehicles for becoming-animal, thus allowing us to reassess the human condition.

Of course, the very idea of the human is defined by our perceptions and conceptions of everything that is not human. Until recent developments in biology and other natural sciences—but also the humanities—started questioning the clear-cut differentiation between human and non-human animals, non-human animals served as the clear choice for the (purportedly) most apparent binary opposite to the human. In this context, Jacques Derrida has observed that "the Animal" is a linguistic construct that refers to "every living thing that is held not to be man" (2008, 31). Humans produce and propagate the difference between human beings and other animals through a variety of discourses. This network of discourses "decides what we recognize and what we don't, what counts and what doesn't; and it also determines the consequences of falling outside the frame" (Wolfe 2013, 6). In this discursive web, the definition of the human generally depends on man's purging of the beast within and without. However, "the taming of men" (Sloterdijk 2009, 15) acknowledges the animal within and without as a key component to defining the human. In other words, by seeking to exorcize the Animal Other, by trying to construct the non-human animal as the binary opposite to the human, the Animal Other is, in fact, inscribed to the human, as the category of the human becomes contingent upon its relationship to the animal.

Video games make this interspecies entanglement explicit, as they rely upon the engagement and interaction with a decidedly constructed version of the animal. I will exemplify this encounter with the non-human by discussing *Bear Simulator* (Farjay Studios 2016), a video game in which—as the title suggests—players control a bear roaming a forest. Reactions to the initial versions of the game, which merely consisted of players acting bear-like and engaging in mundane tasks, such as feeding, drinking, and sleeping, made obvious how uncomfortable the encounter with the non-human Other is when stripped of certain anthropocentric expectations, such as narrative and teleology. Indeed, in early versions of the game, the implied goals were keeping the players' virtual stand-in alive and exploring the virtual world, which, however, were rather easy tasks that offered little gratification and barely any incentives for engaging with the simulated animal in more detail. "Simulation" is key here, though, for *Bear Simulator* offers players a mediated encounter with the animal. As I will suggest, whereas video games may be vehicles for becoming-with-many, the potentials thus opened up often remain unrealized, as they would challenge traditional understandings of the human.

1. Becoming-Animal Avatar

Miroslaw Filiciak has explained that the avatar is "the user's representative in the virtual universe" (2003, 89). As such, the avatar defines "what the players are able to do in the game-world" and "how they do it" (Rouse 2005, 310). To be sure, players' inputs are transformed into gameworld actions, as the pushing of buttons, movement of the mouse, or, more recently, movement of the player's body is kinaesthetically projected onto the avatar, which then performs these actions in the virtual space. The player's resultant feeling of agency in the digital world aids immersion in the gameworld and identification with the player-character. Through this strong bond between the player and the avatar, the player perceives and experiences the virtual world as phenomenologically real, as the differences between the player and avatar fade away. James Newman has accordingly observed that "[t]here is no 'Mario' or 'Sonic' to the player—there is only 'me' in the gameworld" (2004, 134). Jonathan Boulter has similarly diagnosed that "when I am alone with my game console, . . . I assume the role, the identity of the avatar," as "[t]he avatar . . . extends my identity and my space" (2015, 8).

The connection between players and their avatars is thus defined by what Donna Haraway might refer to as "transgressed boundaries" and "potent fusions" (1991, 154), as "[t]he lines separating . . . digital and material worlds dissolve in the space of becoming" (Cremin 2016, 98). In other words, the avatar allows players to simultaneously extend into the game space from the physical space outside it and to occupy the game space, since the avatar remediates the player's agency in the virtual world. The avatar thus "becomes . . . part of the player's 'I can,'" as "[t]he player is . . . re-wired and re-directed . . . through the integrated prosthetic apparatus of controller and . . . avatar" (Klevjer 2012, 27–8). As a result, during gameplay, not only players and virtual environments interact but, maybe more importantly, players and their avatars constantly interact, as well. While playing, the player exists "as a composite of flesh and technology" (Klevjer 2012, 34), becoming something "more" than human.

Arguably, in her book *When Species Meet*, Donna Haraway echoes (or maybe even anticipated) these ideas about video game avatars and their constant interaction with players, claiming that "those who are to be in the world are constituted in . . . interaction. . . . [S]pecies of all kinds, living and not, are consequent on a subject- and object-shaping dance of encounters," which is why "[t]o be one is always to become with many" (2008, 4). Accordingly, players instigate the process of becoming-avatar not by imitation but through composition. Becoming-avatar allows players to enter affective alliances with Others—intimate conjoinings of not only self and machine but also human and other non-human Others.

As a result, the ordinary world of material reality and the extraordinary digital dimension bleed into one another through a process of gradual contagion. "In the lived enactment of gameplay," Jon Dovey and Helen Kennedy have remarked, "there is no player separate to the interface and the game world; there is a fusion of the two into a cyborgian subjectivity" (2006, 109). Indeed, "[p]lay fulfills its purpose only if the player loses himself in play," Gadamer has argued (2013, 107). Importantly, "the player never ceases to feel her real body or the fact that she is holding a controller playing a game" (Owen 2017, loc. 596), but when players lose themselves in a video game, they no longer self-consciously engage in the act of playing.

Crucially, in their musings on Franz Kafka's *Die Verwandlung* (1915), in which the protagonist Gregor Samsa wakes up to find himself transformed into a large insect, Deleuze and Guattari write that human-animal relations are

> a sort of conjunction of two deterritorializations, that which the human imposes on the animal by forcing it to feel or serve the human, but also that which the animal proposes to the human by indicating ways-out or means of escape that the human would never have thought of by himself.
>
> (1986, 35)

Drawing on these ideas, becoming-animal avatar acknowledges that players impose their will on the avatar (while still opening up avenues for becoming-many). Through the interface of the gamepad, mouse, or any other type of controller, players, of course, control their avatars—players effectively tell their avatars where to go to and what to do. However, whereas Bob Rehak thus concluded that the avatar is "an inferior other" (2003, 107), there is an interdependence at work between players and their avatars, as players rely on their avatars to act in the gameworld and to progress in the game. As a result, in a never-ending feedback loop, the avatar opens up ways to proceed for the player. Accordingly, avatars allow players to surpass what they are capable of doing without them.

For example, in *I, Predator!*, depending on whether they play as a prey species or predators, players are tasked to either kill animals or evade predators' attacks. Of course, without their avatars, players could neither kill the virtual animals nor evade the virtual predators' attacks. Significantly, when, for example, playing the "Shark vs. Seal" scenario as the shark, the player does not kill the virtual seals, but neither does the virtual shark (i.e. the avatar)—the *assemblage* consisting of player and shark avatar (in addition to the input device used, computer, etc.) kills the virtual animal. The animal avatar thus emerges as a molecular entity which allows players to exceed the manifold limits of their physical bodies. By becoming-animal avatar, players acquire their avatars' abilities, connecting with lions, sharks, and cheetahs (in the case of the predator scenarios in *I, Predator!*), alongside environmental effects such as (simulated) rainfall and (simulated) wind, creating assemblages that are being composed and collapsed in the moment of play. Indeed, all of these assemblages only exist momentarily, as the video game "only ever comes into being when the game is actually played" (Galloway 2006, 104).

2. Becoming-Bear in *Bear Simulator*

When describing Timothy Treadwell's attempts at connecting with the wild grizzly bears he spent 13 summers with in Alaska, *Grizzly Man*'s (2005) director Werner Herzog remarks, "Treadwell reached out seeking a primordial encounter . . ., as if there was a desire in him to leave the confines of his humanness and bond with bears." Treadwell desperately sought to establish some sort of a link to the grizzlies, tried "to become like the bear," as ecologist Marnie Gaede puts it in the documentary. For anyone conversant in Deleuzo-Guattarian philosophy, Gaede's wording evokes their notion of "becoming-animal," a process they describe as an "unnatural participation" (2004, 265), which "disrupts orders, forms, and substances" (2004, 370). This combination creates a "multiplicity" which transforms the animal and human alike. Deleuze and Guattari suggest that the human body (*any* body, for that matter) is characterized by vectors which branch off into wildly different directions, creating connections with other material and conceptual bodies. Accordingly, the material and immaterial interconnections between human bodies and other bodies, both animate and inanimate, define humans. *Bear Simulator*—that is, both the video game and its surrounding paratexts—taps into these ideas.

Bear Simulator is a rather simple game: "It's a game where you play as a bear and explore stuff," reads the simultaneously intuitive and simplistic tagline on the download platform Steam ("Bear Simulator" 2016). As such, players move their bears through a wilderness area (possibly a national park, but this aspect is not made entirely clear) which seems to be a combination of continental and semiarid steppe climate and which shows traces of human civilization, such as camping sites, vacated buildings, and walls (which, conveniently, tend to serve as borderlines between areas of the forest). Players' actions are rather limited: They can move their bear avatar with different speeds, kill prey species, attack (potential) threats (including a number of boss fights), and engage in other interactions with various objects in the virtual world. The latter interactions primarily revolve around consuming food—from animate creatures, such as maggots and bugs, to burgers and pizza—and various types of liquids and smelling other animals' scent marks, which trigger brief notes informing human players about the messages the bear decodes by sniffing them. These scent marks usually alert players of nearby predators (such as wolf packs) and other—sometimes fantastic—creatures (such as Bigfoot). By consuming various items, players increase or decrease between one and three of the virtual bear's attributes (health, stamina, strength, intelligence, walk, run, swim, dive, jump, and luck). Notably, starvation and dying of thirst do not seem to be deadly threats to the virtual creature, as players need to keep their eyes on two metres only—a blue stamina bar and a red health bar. Similar to the biological imperatives of eating and drinking, reproduction is absent from this "simulation" of a bear's life. In addition, the virtual bears may be "customized" by, among others, having them wear various pieces of headgear, glasses, and even nail polish. Overall, the game is rather easy (albeit sometimes frustratingly difficult because of bad gameplay mechanics), but, in early versions, it lacked objectives other than the implicit goal of basically exploring the entire virtual world.

Upon launching *Bear Simulator*, players find their bear avatar in a cave. The video game includes no introduction of any kind—at first, players inhabit a rather indistinct body in a cave. Accordingly, players have to make sense of the situation they find themselves in. Boulter has suggested that the beginning of a game presents an existentialist moment akin to what Martin Heidegger considered characteristic of a human coming to consciousness in the world—humans' "thrownness" (Heidegger 2010, 131–43, 172–5). As Boulter has explained, "the human is simply thrown into the world . . . with no obvious guidance, . . . having no ready-made philosophical or religious systems with which to make sense of the experience" (2015, 18). Yet before humans can even start to ponder philosophical or religious questions, they must learn to control their bodies, to move, and to navigate the world, so do bears (and other non-human animals) and so do players, as they begin to interact with—and act through—their bear avatars.

Significantly, as soon as players move their bears, an achievement pops up: "Achievement Be a Bear Complete!" The intra-uterine setting of the cave emphasizes the players' symbolic rebirth as something more-than-human. The first-person point of view supports this merging with the bear avatar, as it creates "the illusion of inhabiting the same space as the avatar and see what they see" (Bryce and Rutter 2002, 71). In other words, *Bear Simulator*'s opening moments make explicit how the bear avatar and the player become one—not materially, but phenomenologically and affectively. The human player—no matter the ethnicity, gender, sexual orientation, and so on—becomes-with the bear while playing *Bear Simulator*. Indeed, the allegorical rebirth as something other-than-human, highlighted by the achievement popping up, suggests that the video game opens up a new state of being, as players are thrown into a world that is not their own and a subject position not their own, but which become their own through gameplay.

While the virtual world players are thrown into is unknown, *Bear Simulator* helps players ease into this somewhat strange space by confronting them with a visual configuration they will know. After all, they find themselves inhabiting the body of an animated being whose point of view they share, thus evoking the first-person shooter (FPS). Among video game players, Gerald Vorhees, Josh Call, and Katie Whitlock have remarked, the FPS "is a taken-for-granted form" (2012, 4). This well-established formula includes various elements, chief among them the eponymous first-person perspective, which is intricately connected to movement in the virtual world. Together, these two elements generally suggest "a world that ha[s] similar physical rules to those of 'real life'" (Bryce and Rutter 2002, 68). Another trait of the FPS is the constant presence of a weapon (or, rather, part thereof) on the screen. The weapon's omnipresence, Vorhees, Call, and Whitlock have correctly noted, highlights "the centrality of conflict to gameplay" (2012, 4). *Bear Simulator* conveys this centrality of conflict by allowing players to occupy the virtual body of some*thing* whose vision they share. To be sure, the bear's front legs and paws replace the barrel of a gun, but, in terms of their ludic function, they also serve as weapons, as players quickly discover.

As soon as players start to move their virtual stand-ins, the players' "sense of agency and experience of interactivity" (Morris 2002, 89) deepens the connection between players and their virtual bears. The convergence of vision is amplified by the players' experience of acting in the world through the mediator of the bear avatar. The underlying game mechanics reward players' actions in the virtual world by upgrading the bears' attributes (a mechanic typical of role-playing games), which furthers the connection with the avatar. Crucially, even though the "corporeal body of the player remains in the real world," the "thinking/feeling projected sense of the player's physical presence is extended into the game world" (Owen 2017, loc. 461). The resultant "distributed cognition . . . complicates individual agency" (Hayles 1999, 4), since players cannot act without their bear avatars, and the bear avatars not without their users. As a result, whenever we "use or employ a technology," we are "used by and employed by that technology as well" (Ihde 2002, 137). Accordingly, one could argue that, on a very implicit level, *Bear Simulator* promotes an understanding of the technologically determined lives of both non-human *and* human animals by allowing players to occupy the space of a bear, not simply to *become* a (digital) bear but rather to *become-with* a (digital) bear.

3. Becoming-Non-human in Video Games

Instead of focusing on the meaning potentials and ideas that *Bear Simulator* opened up, players quickly highlighted the game's weaknesses. To keep the background information about *Bear Simulator*'s production and reception to a minimum, suffice it to say that the game was funded through Kickstarter, where nearly 4,000 people pledged more than $100,000. In a tongue-in-cheek manner, the campaign diagnosed "a disturbing trend in games of late" (Farjay Studios 2014)—the dominance of human avatars. As John Farjay, the game's creator, continued, the majority of games "harbor a horrible bias" towards non-human creatures. The dominant system, he suggested, does not "want us playing as bears," but *Bear Simulator* would allow players "to fight back against the tyranny." Initially, the game received largely positive reviews, primarily because players acknowledged the efforts the first-time developer put into the design of the virtual world and the game's humour, which had already been evident in the Kickstarter campaign. However, on 29 February 2016, YouTube icon PewDiePie slammed the game in a video review, which first made the developer implement some updates but eventually led to Farjay's abandonment of the project. Among others, PewDiePie complained, "There's no way this game can be finished. There's nothing to do in the game" (2016). This comment exposes an underlying expectation with which most players approach a video game—performing actions which become meaningful through a (or a number of) pre-defined goal(s). Life does not work this way, however. And a bear's life definitely does not unfold along these human-made vectors of progress.

Since, especially in the wake of PewDiePie's review, a number of players expressed their dissatisfaction with "aimlessly" moving around the world and engaging in "meaningless" tasks, one of the changes Farjay implemented in response to (not only) PewDiePie's criticism was specific objectives. Originally, the game had no stated goals—not even survival as such, as "death" simply leads to the re-spawning of the virtual bear in one of its dens or beds. Players accordingly moved their bear avatars through the virtual environment, thereby exploring the world. Along the way, they looked for food and water, avoided (or confronted) potential dangers, and took naps—things an actual bear does. Here, the aforementioned gameplay mechanic of sniffing at scent marks plays a crucial role, as it represents an action that (potentially) removes the bear avatar from its anthropocentric anchor. Indeed, by acknowledging smell as a communication system humanity shed during the species' path of "progress," *Bear Simulator* exposes a limitation on the part of humans, thereby undermining the notion of humankind's undisputed superiority to other life forms on the planet. Problematically, however, the information the virtual bear perceives and receives through its senses is transcoded into the English language. To be sure, humans are trapped in a process whereby any attempt at understanding the non-human entails transcoding a non-human creature's experience into human perceptions and conceptions of the world. When not even trying to comprehend non-human creatures, "we will lose contact with a large part of our world. If I cannot say that a dog is sad, what can I say that it is? In a sense, without anthropomorphism we are unable to comprehend and represent the present of an animal," Erica Fudge has remarked (2002, 76). While she is right to point out that without approaching animals from such an anthropocentric perspective, "the only relation we can have with animals is a very distant, and perhaps mechanistic one" (Fudge 2002, 76), at the end of the day, "[l]anguage is one of the tools we use to . . . explain and master nature" (Sullivan 2013, 38).

The implementation of explicit objectives shifted the game's focus even further in the direction anticipated by the interspecies translation of the scent marks. In later versions of the game, the virtual bear comes across posters and notes plastered onto trees, rocks, and other objects, which sketch the main tasks in writing. Players can find one of these notes in the cave in which the game opens, outlining the main "bear goals":

1. Find and eat everything in each region.
2. Become smarter than the average bear.
3. Stay away from further regions until I bulk up.
4. Discover mystery.

(Farjay Studios 2016)

Implicitly, these objectives had been part of *Bear Simulator* from day one, embedded into the virtual world. But the inclusion of these tasks in written form translates the "bear-like" discovery of "tasks" by engaging with the environment into a list that may be ticked off one by one, thereby imposing a narrative of progress onto the game and transforming the chaos of nature into human-made order. Later versions of the game, in particular, hence reveal humankind's "pathological belief in our ability to control the . . . natural world" (Williston 2015, 35).

I would argue that *Bear Simulator* and the changes implemented after its original release thus highlight three important ideas connected to human—animal relationships in the context of digital media that I would like to conclude this chapter with.

First, *Bear Simulator*, as one response to Farjay's announcement that he would abandon the project stressed, "is just that"—a bear simulator, which means that "it simulates an actual animal" (Brinck 2016). Of course, in discourses of cultural studies and media studies, "simulation" is inseparably tied to the name of Jean Baudrillard. Interestingly, video game scholars have repeatedly highlighted a purportedly clear difference between Baudrillard's notion of "simulation" and what the term means in the context of video games. For example, Noah Wardrip-Fruin and Pat Harrigan have stressed that "[t]he simulation discussed by [Baudrillard] is a cultural phenomenon, not a computational one—and as such is fully existent in old media as well as new" (2004, 71). However, as James Campbell has more recently pointed out, "[I]t is precisely because Baudrillard's version of simulation is a cultural rather than a merely digital phenomenon that it has important implications for computer game studies" (2008, 187).

To be sure, Baudrillard suggested that the operating principle of simulation is "substituting the signs of the real for the real" (1994, 2). As a result, the difference between what is real and what is not vanishes, as any discursively created distinction between reality and fiction becomes artificially constructed, leading to a never-ending chain of signification which not simply replaces reality but even makes its recovery impossible. In this way, the symbolic realm "subtly detach[es]" objects "from their primal reality" (Baudrillard 2009, 24–5), as "global technology" (and the media apparatus fuelled by this technology) "denature[s] everything" (Haraway 1992, 295). By trying to simulate the lives of wild animals, *Bear Simulator* replaces actual animals with virtual stand-ins. The video game thus simultaneously highlights and contributes to wild animals' disappearance from the everyday experiences of Westerners. In this context, Akira Mizuta Lippit has concluded that it could almost be said that "[a]nimals found a proper habitat . . . in the recording devices of the technological media" (2000, loc. 400), which have allowed humankind to transform animal lives into data in an attempt to store and preserve them. Wild animals have, accordingly, taken on a spectral form in contemporary Western culture. They are a "non-present present" (Derrida 2006, 5), as they are simultaneously (omni)present in the media and absent from Westerners' lives.

While, in a way, the bear avatar replaces (and displaces) real animals, it (somewhat paradoxically) allows players to affectively engage with the non-human. This second concluding remark is connected to *Bear Simulator*'s practical desertion of teleology and the related abandonment of traditional narrative principles. Indeed, *Bear Simulator* does not try to tell a well-made story featuring a beginning, middle, and ending but rather throws players into the middle of a bear's life and makes them engage in ordinary and rather mundane minutiae of everyday life, with no attempt at establishing causal relations or "logic." (To be sure, this does not mean that players might not create narrative links themselves.) Tellingly, Gilles Deleuze has maintained that once "the detour and boredom of conveying a story" has been overcome, affect remains, which "is transmitted directly" (2003, 36). The affective connection established between the bear avatar and the player in combination with the recognition that the player needs the avatar in order to act in the virtual world as much as the avatar needs the player in order to act in the virtual world suggest that "[t]he animals make demands on the humans and their technologies to precisely the same degree that humans make demands on the animals" (Haraway 2008, 263).

This mutual dependence leads to the third—and final—point on my list: the perhaps "dangerous possibilities" (Haraway 1991, 154) opened up by this relationship between human and non-human. The bear avatar, as I have outlined earlier, is neither entirely dependent on the players nor fully independent from them; the avatar is neither powerless nor autonomous—but neither are the human players. This interdependence between the human and the non-human highlights their interrelation. When becoming-bear avatar, players dissociate themselves from parts of their humanity and temporarily transform into a hybrid, post-human creature entangled with human, bear, avatar, and computer. This acknowledgement of the non-human's importance to the (post)human has the potential to upset the established anthropocentric worldview, in which humankind sits both atop and apart from the rest of the world, and would allow for a more comprehensive understanding of the world. However, the players' reactions to the post-human subtexts and potentials of the original versions of the game, in particular, show that such an interpretation would be too optimistic. Yes, video games offer means to "enhance humans beyond broadly accepted natural and cultural limits" (Sharon 2014, 17), to transcend the human body's biological limits through technology, and to embrace "the radical otherness of other life forms" (Sabloff 2001, 149), but human beings sit simply too comfortably in their imagined role as coherent and cohesive entities separate and distinct from the "natural" world. Humans find reassurance by picturing themselves disentangled from the non-human domain. This worldview allows the human species to imagine itself as the dominant species on this planet, and it allows *Homo sapiens* to exploit the planet and act as if there were an alternative Earth somewhere where the living creatures and inanimate matters of this planet could continue to exist when our current homeworld will have become the inhabitable place it is slated to turn into sooner rather than later.

References

Baudrillard, Jean (1994). "The Precession of Simulacra." In *Simulacra and Simulation*, translated by Sheila Faria Glaser, 1–42. Ann Arbor: University of Michigan Press.

Baudrillard, Jean (2009). "On Disappearance." In *Fatal Theories*, edited by David B. Clarke, Marcus A. Doel, William Merrin, and Richard G. Smith, 24–9. Abingdon: Routledge.

"Bear Simulator" (2016). *Steam*. Accessed August 15, 2017. http://store.steampowered.com/app/395850/Bear_Simulator/.
Beggs, Kevin L., et al. (Producers), and Werner Herzog (Director). 2005. *Grizzly Man* [Motion Picture]. USA: Real Big Production.
Boulter, Jonathan (2015). *Parables of the Posthuman: Digital Realities, Gaming, and the Player Experience*. Detroit: Wayne State University Press.
Brinck, Kevin (2016). Comment to "Bear Simulator: Update No. 45." *Kickstarter*. Accessed August 15, 2017. www.kickstarter.com/projects/1330754003/bear-simulator/posts/1510978?cursor=12824712#comment-12824711.
Bryce, Jo, and Rutter, Jason (2002). "Spectacle of the Deathmatch: Character and Narrative in First-Person Shooters." In *Screen Play: Cinema/Videogames/Interfaces*, edited by Geoff King and Tanya Krzywinska, 66–80. London: Wallflower.
Campbell, James (2008). "Just Less Than Total War: Simulating World War II as Ludic Nostalgia." In *Playing the Past: History and Nostalgia in Video Games*, edited by Zach Whalen and Laurie N. Taylor, 183–200. Nashville, TN: Vanderbilt University Press.
Cremin, Colin (2016). *Exploring Videogames with Deleuze and Guattari: Towards an Affective Theory of Form*. Abingdon: Routledge.
Deleuze, Gilles (2003). *Francis Bacon: The Logic of Sensation*. Translated by Daniel W. Smith. London: Continuum.
Deleuze, Gilles, and Guattari, Félix (1986). *Kafka: Towards a Minor Literature*. Translated by Dana Polan. Minneapolis: University of Minnesota Press.
Deleuze, Gilles, and Guattari, Félix (2004). *A Thousand Plateaus: Capitalism and Schizophrenia*. Translated by Brian Massumi. Minneapolis: University of Minnesota Press.
Derrida, Jacques (2006). *Specters of Marx: The State of Debt, the Work of Mourning, and the New International*. Translated by Peggy Kamuf. New York: Routledge.
Derrida, Jacques (2008). *The Animal That Therefore I Am*. Edited by Marie-Louise Mallet. Translated by David Wills. New York: Fordham University Press.
Dovey, Jon, and Kennedy, Helen (2006). *Game Cultures: Computer Games as New Media*. Maidenhead: Open University Press.
Farjay Studios (2014). "Bear Simulator: About This Project." *Kickstarter*. Accessed August 15, 2017. www.kickstarter.com/projects/1330754003/bear-simulator/description.
Farjay Studios (2016). *Bear Simulator* [Video Game]. Bellevue, WA: Steam.
Filiciak, Miroslaw (2003). "Hyperidentities: Postmodern Identity Patterns in Massively Multiplayers Online Role-Playing Games." In *The Video Game Theory Reader*, edited by Mark J.P. Wolf and Bernard Perron, 87–102. New York: Routledge.
Fudge, Erica (2002). *Animal*. London: Reaktion Books.
Gadamer, Hans-Georg (2013). *Truth and Method*. Translated by Joel Weinsheimer and Donald G. Marshall. London: Bloomsbury.
Galloway, Alexander R. (2006). *Gaming: Essays on Algorithmic Culture*. Minneapolis: University of Minnesota Press.
Haraway, Donna J. (1991). *Simians, Cyborgs, and Women: The Reinvention of Nature*. New York: Routledge.

Haraway, Donna J. (1992). "The Promises of Monsters." In *Cultural Studies*, edited by Lawrence Grossberg, Cary Nelson, and Paula A. Treichler, 295–337. New York: Routledge.

Haraway, Donna J. (2008). *When Species Meet*. Minneapolis: University of Minnesota Press.

Hayles, N. Katherine (1999). *How We Became Posthuman: Virtual Bodies in Cybernetics, Literature, and Informatics*. Chicago: University of Chicago Press.

Heidegger, Martin (2010). *Being and Time*. Translated by Joan Stambaugh. Albany: State University of New York Press.

Ihde, Don (2002). *Bodies in Technology*. Minneapolis: University of Minnesota Press.

"I, Predator" (2009). *Kuma Games*. Accessed August 15, 2017. www.ipredator.tv/.

Jameson, Fredric (1991). *Postmodernism, or, the Cultural Logic of Late Capitalism*. Durham, NC: Duke University Press.

Klevjer, Rune (2012). "Enter the Avatar: The Phenomenology of Prosthetic Telepresence in Computer Games." In *The Philosophy of Computer Games*, edited by Hallvard Fossheim, Tarjei Mandt Larsen, and John Richard Sageng, 17–38. Leiden: Springer.

Kuma Games (2011). *I, Predator* [Video Game]. New York: Kuma Games.

Lindberg, Adam (2017). "To Torture the Torturer: Videogame Violence and the Question of Humanism." *TRACE: Journal of Writing, Media, and Ecology* 1: n. pag.

Lippit, Akira Mizuta (2000). *Electric Animal: Toward a Rhetoric of Wildlife*. Kindle Edition. Minneapolis: University of Minnesota Press.

Morris, Sue (2002). "First-Person Shooters: A Game Apparatus." In *ScreenPlay: Cinema/Videogames/Interfaces*, edited by Geoff King and Tanya Krzywinska, 81–97. London: Wallflower.

Newman, James (2004). *Videogames*. Abingdon: Routledge.

Owen, David (2017). *Player and Avatar: The Affective Potential of Video Games*. Jefferson, NC: McFarland. Kindle edition.

PewDiePie (2016). *Best Game 2016!!!!!!!!!!!!!!!!!!!!!!!!!!!!!!!!!!!—(Bear Simulator)*. YouTube. Accessed August 15, 2017. www.youtube.com/watch?v=hbVUF5b307s.

Rehak, Bob (2003). "Playing at Being: Psychoanalysis and the Avatar." In *The Video Game Theory Reader*, edited by Mark J.P. Wolf and Bernard Perron, 103–27. New York: Routledge.

Rouse, Richard, III (2005). *Game Design: Theory & Practice*. 2nd Edition. Plano, TX: Wordware Publishing.

Sabloff, Annabelle (2001). *Reordering the Natural World: Humans and Animals in the City*. Toronto: University of Toronto Press.

Sharon, Tamar (2014). *Human Nature in an Age of Biotechnology: The Case for Mediated Posthumanism*. Dordrecht: Springer.

Sloterdijk, Peter (2009). "*Rules for the Human Zoo*: A Response to the *Letter on Humanism*." *Environment and Planning D: Society and Space*, no. 27: 12–28.

Sullivan, Rosemary (2013). "*La forêt* or the Wilderness as Myth." In *Greening the Maple: Canadian Ecocriticism in Context*, edited by Ella Soper and Nicholas Bradley, 31–42. Calgary: University of Calgary Press.

Ulmer, Gregory L. (2012). *Avatar Emergency*. Anderson, SC: Parlor Press.

Vorhees, Gerald, Call, Josh, and Whitlock, Katie (2012). "Introduction: Things That Go Boom: From Guns to Griefing." In *Guns, Grenades, and Grunts: First-Person Shooter Games*, edited by Gerald A. Vorhees, Josh Call, and Katie Whitlock, 1–21. New York: Continuum.

Wardrip-Fruin, Noah, and Harrigan, Pat (2004). "Critical Simulation." In *First Person: New Media as Story, Performance, and Game*, edited by Noah Wardrip-Fruin and Pat Harrigan, 71–2. Cambridge, MA: MIT Press.

Williston, Byron (2015). *The Anthropocene Project: Virtue in the Age of Climate Change*. New York: Oxford University Press.

Wolfe, Cary (2013). *Before the Law: Humans and Other Animals in a Biopolitical Frame*. Chicago: University of Chicago Press.

Biographical Statements

Éric Baratay is a member of the Institut Universitaire de France and professor of contemporary history at the University of Lyon. He developed a human history of animals by studying representations (*L'Église et l'Animal*, 1996, *Des Bêtes et des Dieux*, 2015) and human practices (*La Corrida*, 1995; *Zoos*, 1998), and then the condition of animals (*Et l'homme créa l'animal*, 2003) and their presence in society (*Bêtes de somme*, 2011). Since then, he has been working on experiences and points of view in order to build an animal history of animals (*Le Point de vue animal. Une autre version de l'histoire*, 2012; *Bêtes des tranchées. Des vécus oubliés*, 2013); *Biographies animales. Des vies retrouvées*, 2017), an historical eth*(n)*ology of animals.

Eva Voldřichová Beránková studied French philology at the Faculty of Arts, Charles University (Prague), and Slavonic literatures at the Université Paris IV-Sorbonne as part of a joint doctorate. Since 2013, she has been an associate professor in the Institute of Romance Studies at the Faculty of Arts (Charles University) and head of its French department. She specializes in French literature from the turn of the nineteenth and twentieth centuries, literary theory, the history of film, and the Francophone Quebec novel. She has published two monographs (*The Hidden Dostoevskian Face of Albert Camus*, 2002; *"Let Us Make Man in Our Image": Pygmalion, Golem and the Automaton as Three Versions of the Myth of Artificial Creation*, 2012), four collective monographs, and many chapters of books and articles in Czech and international scientific journals. Her most recent research focuses on the reception in France of Hegel's *Aesthetics*.

Concepción Cortés Zulueta is a Juan de la Cierva post-doc fellow at Universidad de Málaga, Spain. She received her PhD in art history from Universidad Autónoma de Madrid in 2016. Her research, interdisciplinary in scope, focuses on the presence and agency of non-human animals in contemporary art and visual culture from the 1960s to the beginning of the twenty-first century, approaching their roles as artworks, viewers, interlocutors, and artists inside and beyond the contemporary art gallery. She has been a temporary researcher at La Sapienza, Università di Roma; the National Art Library (Victoria & Albert Museum, London), the New Zealand Centre for Human-Animal Studies (University of Canterbury, Christchurch), and the Massachusetts Institute of Technology.

Kári Driscoll is assistant professor of comparative literature at Utrecht University in the Netherlands. Along with Eva Hoffman, he is Co-editor of *What Is Zoopoetics?—Texts, Bodies, Entanglement* (Palgrave Macmillan 2018) and, with Susanne C. Knittel, of "Memory after Humanism," a special issue of *Parallax* (2017). His current research project, funded by the Dutch Research Council (NWO), is entitled "Reading Zoos in the Age of the Anthropocene."

Michael Fuchs is a postdoctoral researcher at the Center for Inter-American Studies at the University of Graz in Austria. He has co-edited six essay collections, among them a special issue of the *European Journal of American Studies* on animals on American television (2018) and *Intermedia Games—Games Inter Media: Video Games and Intermediality* (2019). In addition, he has authored and co-authored more than 50 journal articles and book chapters on topics including video games, American television, horror cinema, science fiction, and comics. One of his ongoing monograph projects explores the role of monstrous animals in American culture.

Maria P. Gindhart is associate professor of art history at Georgia State University, USA, where she currently serves as associate dean of the College of the Arts. Her previous work on images of prehistoric humans has appeared in *The Journal of the History of Collections*, *Nineteenth-Century Art Worldwide*, and *Visual Resources*. Her subsequent scholarship on the Jardin des Plantes, and in particular the sculpture displayed there, has been published in *Material Culture Review* and *The Heyday of Sculpture 1850–1880* (Paris, 2018). Her current research includes a book-length project on non-human animals, both real and represented, at the 1931 International Colonial Exposition in Paris.

Anita Jarzyna, PhD at University of Łódź (Poland), is a literary researcher. Her main field of interests are poetry, animal studies, and Holocaust studies. She is the author of three books and several essays. Her latest book is entitled *Post-koiné. Studies of Non-Anthropocentric Poetic Languages* (2019). She is also a member of the editorial board of the journal *Narrations of the Shoah*.

Michał Krzykawski, is associate professor and director of the Centre for Critical Technology Studies at the University of Silesia in Katowice, Poland, and member of the Internation/Geneva2020 research group founded on the initiative of Bernard Stiegler. He has written extensively on contemporary French philosophy. He is the author of two monographs: *L'effet-Bataille. De la littérature d'excès à l'écriture. Un texte-lecture* (Katowice, 2011) and (in Polish) *The Other and the Common. Thirty-Five Years of French Philosophy* (Warszawa, 2017). His two books (in Polish), "The Condition of the Future. A New Report On Knowledge" and "The Soul, the Spirit, and Technics. A Philosophical Essay for the Youth," are forthcoming.

Chiara Mengozzi is a researcher at Charles University in Prague (Faculty of Arts), where she teaches literary theory, as well as Italian and French contemporary literatures. She is also Associated Researcher at CEFRES in Prague (French Research Center in Social Sciences). Her research focuses on animal studies, ecocriticism, climate fiction, postcolonial and migrant literatures, world literature, and translation and reception theory. She is the author of an Italian monograph entitled *Narrazioni contese. Vent'anni di scritture italiane della migrazione* (Carocci, 2013). She is currently investigating the relation between science and literature in the context of climate change.

Quentin Montagne is the author of a doctoral thesis specializing in arts, *The Aquarium: Vision and Representation of the Underwater Worlds. A Device for Exhibition at the Crossroads of Art and Science*, under the direction of Christophe Viart (University of Rennes 2, France). As a research fellow at the laboratory Practices and Theory of Contemporary Art in this institution, he regularly participates in seminars and study days. Also graduate of EESAB (European Superior School of Art of Brittany, Rennes), he combines theoretical research and artistic production, which he regularly testifies through exhibitions and editions.

Rachael L. Pasierowska is a dual PhD candidate in history at Rice University, USA, and the Universidade Estadual de Campinas, Brazil. She received her BA in combined modern languages at the University of Portsmouth, her MSc in slavery and forced labour studies at the University of Edinburgh, and her MA in history at Rice University. She is the author of two award-winning articles "Screech owls allus holler 'round the house before death': Birds and the Souls of Black Folk in the Antebellum South" and "Up from Childhood: When African-American Enslaved Children Learned of their Servile Status." Her forthcoming dissertation is a study of the intricate relations between slaves and animals in the nineteenth-century Atlantic World, with a focus on the slave societies of Brazil, Cuba, and the US South. She has received prestigious fellowships from multiple archives to research her dissertation, including the American Antiquarian Society, the American Philosophical Society, the Huntington Library, the Smithsonian Institution, and the University of Miami.

Jonathan Pollock is professor of English and comparative Literature at the University of Perpignan-Via Domitia, France. His publications include the following books or edited volumes: *Qu'est-ce que l'humour?* (2001), *Le Moine (de Lewis) d'Antonin Artaud* (2002), *Le Rire du Mômo: Antonin Artaud et la littérature anglo-américaine* (2002), *Invectives* (2005), *Hétérologies* (2006), *Déclinaisons. Le Naturalisme poétique de Lucrèce à Lacan* (2010), *Pratiques du hasard* (2012), *Lire* Les Cantos *d'Ezra Pound* (2014), *Corps, Poésie, Esthétique* (2016).

Violette Pouillard is a postdoctoral research fellow of the Research Foundation-Flanders (Ghent University) and of the National Fund for Scientific Research (Free University of Brussels). She has published a book and several articles on the history of zoos and the (post) colonial management of wildlife (*Histoire des zoos par les animaux*, Champ Vallon, 2019). Her current research project, *Living With(out) Elephants*, further explores the management of wildlife in both the Belgian Congo and British Uganda.

Anne Simon studied at the École Normale Supérieure and Paris Sorbonne, France. She is currently a research director at the Centre National de la Recherche Française and a member of the École des Hautes Études en Sciences Sociales (Paris), where she leads the Zoopoetics Project "Animots" and the "Pôle Proust." Her research focuses on relationships between philosophy and literature, and on literary animal studies. Her main publications are *Les Femmes à l'épreuve du nouvel ordre moral* (with C. Détrez, Seuil, 2006), *Proust ou le réel retrouvé* (PUF, 2000), *Proust e la filosofia contemporanea* (Solfanelli, 2013), *Trafics de Proust* (Hermann, 2016), and *La Rumeur des distances traversées* (Classiques Garnier, 2018). She has also co-edited 15 journal issues, including "Facing animals," *L'Esprit créateur* (2011); "Human-Animal," *Contemporary French and Francophone Studies* (2012); "Écopoétiques", *Fixxion* (2015) and "Zoopoétique," *Revue des Sciences humaines* (2017).

Kari Weil is university professor of letters at Wesleyan University, USA. She is the author of *Precarious Partners: Horses and Their Humans in Nineteenth-Century France* (forthcoming from the University of Chicago Press), *Thinking Animals: Why Animal Studies Now* (Columbia UP, 2012) and numerous essays dealing with human-animal relations, feminist theory, literary representations of animal otherness, and gender. She is starting a new project that will pursue connections between theories of animal magnetism in the eighteenth and nineteenth centuries and contemporary notions of affective influence and biosemiotics.

Index

For Product Safety Concerns and Information please contact our EU
representative GPSR@taylorandfrancis.com
Taylor & Francis Verlag GmbH, Kaufingerstraße 24, 80331 München, Germany

www.ingramcontent.com/pod-product-compliance
Ingram Content Group UK Ltd.
Pitfield, Milton Keynes, MK11 3LW, UK
UKHW022205190926
475702UK00003B/207
9780367504441